mente deciso di creare un'auto perfetta."

Lamborghini

MURCIÉLAGO

THILLAINATHAN PATHMANATHAN
Foreword by TONINO LAMBORGHINI

Other great books from Veloce:

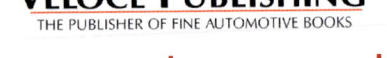

EARTHWORLD EXPANDING HORIZONS

www.veloce.co.uk

First published in October 2018 by Veloce Publishing Limited, Veloce House, Parkway Farm Business Park, Middle Farm Way, Poundbury, Dorchester DT1 3AR, England. Tel +44 (0)1305 260068 / Fax 01305 250479 / e-mail info@veloce.co.uk / web www.veloce.co.uk or www.velocebooks.com. ISBN: 978-1-845849-22-1 UPC: 6-36847-04922-5. © Copyright 2018 Thillainathan Pathmanathan and Veloce Publishing. All rights reserved. With the exception of quoting brief passages for the purpose of review, no part of this publication may be recorded, reproduced or transmitted by any means, including photocopying, without the written permission of Veloce Publishing Ltd. Throughout this book logos, model names and designations, etc, have been used for the purposes of identification, illustration and decoration. Such names are the property of the trademark holder as this is not an official publication. Readers with ideas for automotive books, or books on other transport or related hobby subjects, are invited to write to the editorial director of Veloce Publishing at the above address. British Library Cataloguing in Publication Data – A catalogue record for this book is available from the British Library. Typesetting, design and page make-up all by Veloce Publishing Ltd on Apple Mac. Printed in India by Parksons Graphics.

Lamborghini
MUR
CIÉL
AGO

THILLAINATHAN PATHMANATHAN

Foreword by TONINO LAMBORGHINI

CONTENTS

FOREWORD BY TONINO LAMBORGHINI

My father has left a great legacy for all of us.
His entire life was full of passion and he has expressed his enthusiasm in all the things he has created and conceived. Following in his footsteps, every day in all of my activities and businesses, I try to be inspired by his innovative spirit and great determination that he has always transmitted to me. I think that his creativity and strong character could inspire anyone who seeks to challenge life and stand out, reaching beauty and perfection with an unconventional and uncompromising spirit.

INTRODUCTION

The Murciélago stands proud in the Pantheon of Lamborghini supercars.

Just as the Pantheon (temple of all the gods) was commissioned by Marcus Agrippa and completed by the Emperor Hadrian, the Murciélago was, in a sense, commissioned by none other than Ferruccio Lamborghini, and completed by the Volkswagen-Audi conglomerate.

One of Ferruccio Lamborghini's final acts as the sole owner of Automobili Ferruccio Lamborghini SpA was to sign off on the production of the seminal and revolutionary Countach. For two decades the Countach was synonymous with Lamborghini, and similarly, Lamborghini meant the Countach.

The Murciélago is a direct descendent of the Countach, as evidenced by its spaceframe chassis, its Bizzarrini-derived engine, its south-north engine-gearbox orientation, its genuinely vertical-opening guillotine doors, and – of course – its spectacular wedge silhouette.

The Murciélago reigned supreme for the best part of a decade, and its demise marked the end of the spaceframe and Bizzarrini era at Lamborghini.

There is little doubt that the Murciélago fully deserves its seat (or, more correctly, its couch) at the lectisternium feast of the Lamborghini gods.

... a most unusual car ...
– Toad, 'The Wind in the Willows'

Dedication: To my wife, Dr Anne Christina Reck, who made this book, and so many other things, possible, and to my parents, Siva and Path, without whom, nothing at all would have been possible.

ACKNOWLEDGEMENTS

The reality is that while I spend far too much time on cars, and had been intending to write about the Countach for almost two decades, this Murciélago book – my first attempt at automotive writing – is in existence today solely because of my wife, Dr Anne Christina Reck. When the waters got choppy, and I was in despair or tempted to abandon ship, it was her calm persistence and steadfast encouragement that kept the gondola on course. Thank you also for providing the subject material.

I am also immensely grateful to Rod Grainger of Veloce Publishing for having the courage to take on an enthusiastic, but virginal, author. Jeff 'Jai' Danton was presented with a jumble of words and a tangle of photos and his expertise in design and layout speaks for itself. Jeff also wrote most of the captions, although the responsibility for its accuracy remains all mine. Joe Russell did the editing – thank you. Kevin Quinn, Kevin Atkins and Sian Pettit of Veloce were all invaluable in fine-tuning this book into a form suitable for publication.

Cav. Comm. Dott. Ing. Tonino Lamborghini has been very generous with his time, with writing the foreword for this book, and with providing images from the Museum's archives – 'FERRUCCIO LAMBORGHINI MUSEUM, FUNO DI ARGELATO, HISTORICAL ARCHIVE.' I believe that this is the first stand-alone book on the Murciélago, the car that is the last, true, direct descendent of the eternal Countach. One of Ferruccio Lamborghini's last acts as sole Founder-Owner of Automobili Ferruccio Lamborghini SpA. was to sign off the Countach for production. It is therefore fitting, and an honour, to have some of Ferruccio's blood and spirit in this book, in the form of Tonino Lamborghini. It would also be impossible not to mention Gaetano Tassinari and Dr Elena Grandi of the Museo Ferruccio Lamborghini in Bologna who have been so very helpful, so I am taking the liberty of disregarding their pleas to ignore them in the acknowledgements.

This book has been a collaborative project, and I have been both lucky and privileged to have on-board established Murciélago experts like Ed Bolian, Mike Pullen, Roberto Grimaldi, Ian Hunt, Simon George and Bob Forstner. I would also like to thank Craig Johnson for imparting some of his engineering expertise through his clear explanation of the Murciélago's mechanicals.

Jim Holder of *Autocar*, and Stuart Gallagher, Ryan Chambers and Nick Trott of *Evo* magazine have been kind and encouraging throughout this enterprise, and I would like to thank them for allowing me to use excerpts from *Autocar* and *Evo* in this book.

The Beaulieu National Motor Museum in the United Kingdom is the second home for many of my cars, and I am very grateful to Jonathan Day, Tim David Wood, Russel Bowman and Stephen Munn for the use of the wonderful images taken in the Beaulieu photographic studios.

Ian Hunt – an expert photographer – has been hugely generous in sharing his photographs and contributing to the text, and needs specific acknowledgement. Similarly, John Zuberek and Roy Cats of Cats Exotics sent many, many beautiful photographs of rare Murciélagos from the United States. Thank you also to Celia Chester and Christel Munster for the Valencian photos.

Gerald Kahlke of Lamborghini S.p.A was generous in letting me use drawings and pictures from the factory owners manual.

Sheikh Amari of Amari Supercars, Ian Kershaw and Gary Tolson at Kaaimanns International Supercars, James Huntley and Paul Burrows at SuperVettura Supercars, and Robert and Vanessa Forstner of Bob Forstner Park Lane are all well-established and highly-respected supercar dealers, through whose hands many Murciélagos have passed; thank you all for your wonderful support and the images.

Alexander Sinclair (with photo credits to Jimmy Lelong and Graeme Franklin), Angelo Savoia, Jason Toms, John Rutter, Vince Finaldi (with photo credits to Ben Bertucci), and Maurice Rizzuto all have two things in common: beautiful Murciélagos, and genuine enthusiasm for their Raging Bulls. I am hugely grateful to you for sending me photographs and text descriptions of your cars for use in this book.

Thank you again everyone – the truth is that this book would not be what it is, or possibly not in existence at all, without all your enthusiasm, participation and generosity.

FERRUCCIO LAMBORGHINI AND THE GENESIS OF AN ICONIC MARQUE

*V*alentino Balboni, the Lamborghini factory's legendary test driver, has described Ferruccio Lamborghini as "a simple man, a simple farmer." The founding principle of Automobili Ferruccio Lamborghini SpA was certainly simple and clear in its intention, albeit immensely ambitious and difficult to achieve in practice: to build the world's best Gran Turismo car.

Implicit in this statement is the fact that Lamborghini set out to build road cars, and not racing or quasi-racing cars. Surrounded by Fiat, Alfa Romeo, Maserati and Ferrari – all long-established, and each with a grand history in motor racing – Ferruccio Lamborghini felt that his resources could be best utilised, and his objectives best met, by actively avoiding any involvement in racing.

This is not to say that Ferruccio had a distinctive or inherent dislike of racing. On the contrary, one of his early projects was to modify an old Fiat Topolino into an open top 750cc two seater, with which he entered the 1948 Mille Miglia.

The Fiat 500 Topolino (or 'little mouse') was produced from 1936 to 1955, and was one of the smallest cars in worldwide production at that time. It had a 569cc, four-cylinder, side-valve, water-cooled engine, which was said to produce 13bhp. Lamborghini modified this engine by casting his own cylinder head in bronze, with overhead valves, and attached it to a bored-out 750cc cylinder block.

The Mille Miglia was an open road endurance race, which first took place in 1927. Six years earlier, in 1921, the Italian Grand Prix was moved from Brescia to Monza. In response to this, the young Count Aymo Maggi, from Brescia, together with Franco Mazzotti and a few other well-heeled friends, set up a 1000-mile, figure-of-eight course, running from Viale Venezia in Brescia to Rome and back.

Having previously won the 1926 Formula Libre Rome Grand Prix in a Bugatti, Count Aymo Maggi took part in the inaugural Mille Miglia of 1927 driving an Isotta Fraschini 8A SS, and achieving sixth place.

This original Mille Miglia ran a total of 24 times, from 1927 to 1957, with an interruption for the Second World War.

The 1948 Mille Miglia started at midnight in a very heavy downpour. Tazio Nuvolari was a last minute addition to the Ferrari team, and put in his typical superhuman effort, going as far as using a bag of lemons as a cushion after his seat went adrift. However, he was forced into retirement after damaging his rear suspension at Leghorn, and thereby gifting victory to his team-mate Biondetti, who completed the

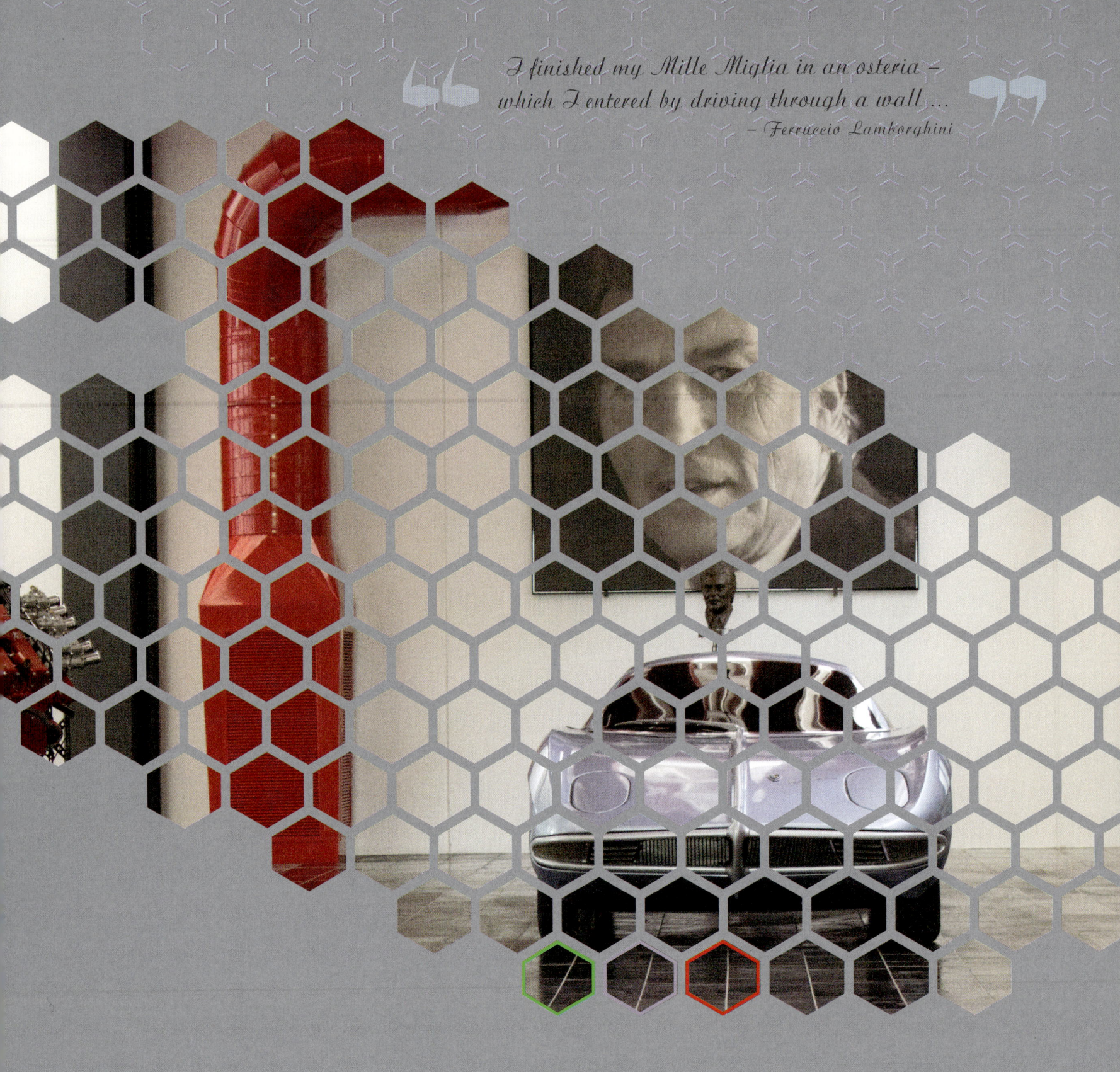

*I finished my Mille Miglia in an osteria –
which I entered by driving through a wall ...*
– Ferruccio Lamborghini

course in 15 hours 5 minutes and 44 seconds at a winning speed of 75.76mph.

In a similar fashion to Nuvolari, Ferruccio Lamborghini bravely competed in his modified Topolino for 700 miles against the might of Ferrari, Fiat, Cisitalia and Alfa Romeo, until he crashed into a roadside restaurant in the commune of Fiano, in the province of Turin.

He emerged unhurt, and later said of the race, "I finished my Mille Miglia in an osteria [tavern], which I entered by driving through a wall."

Ferruccio Lamborghini's enthusiasm for racing may have evaporated following this accident, but it was also hard-headed financial acumen that led Ferruccio to avoid any further involvement in racing.

Over the decades there have been only a few motor racing programmes that Lamborghini has participated in, and those largely insignificant, particularly compared to its Modenese neighbour, Ferrari. A notable exception is that Lamborghini supplied the V12 L3512 engine, which featured an unusual 80-degree cylinder angle, to various Formula One teams between 1989 and 1993, including Larrousse, Lotus, Ligier and Minardi. With the Murciélago, Lamborghini's participation in motor sports was limited to the R-GT and its derivatives.

Lamborghini SpA started off with the sole objective of building road cars, and to appreciate this fact is to better understand how – and why – it has been able to evolve into a manufacturer with the ability to consistently design and build some of the world's most exotic, and most iconic, high-performance cars of the last six decades.

HISTORY

At 9:30pm on the warm summer evening of Friday, 14th July 2006, a statue to commemorate Ferruccio Lamborghini was unveiled in his hometown of Renazzo di Cento.

Cast in the same bronze material as his original Topolino overhead-valve cylinder head, and sculpted by Salvatore Amelio, the composition commemorates Lamborghini's contribution to the industrial and commercial success of the town, and the wider Emilian region.

The three-metre-tall structure has a triangular tip – the symbol of freedom and ingenuity – with a tractor on its right, the Miura on its left, and a bull at the centre, as an expression of power connected to Mother Earth.

The symbol of the raging bull is inextricably connected to every Lamborghini supercar, and adorns the bonnet of each and every one of them. To car enthusiasts, the mere sight of the golden raging bull, set against a black triangular background, is guaranteed to set the heart aflutter. When commissioning the raging bull symbol, Ferruccio is said to have told the designer, "I want a bull with big balls."

Ferruccio Elio Arturo Lamborghini was born on 28th April 1916 in the rural village of Renazzo di Cento, which lies about 15 miles north of Bologna.

This was a time and a place when and where significant importance was given to a person's birth sign, and in Ferruccio's case, he appears to embody the supposed characteristics exhibited by those born under his astrological sign of Taurus: hard-working, tenacious, aggressive, goal-orientated and successful.

The name Renazzo derives from the River Rhine, and Ferruccio Lamborghini's forefathers had farmed in this area for generations. Renazzo's claim to fame was that a meteor had disintegrated in the air above it in January 1824, and three fragments were later recovered from the surrounding farmland. That, of course, was before Ferruccio made Renazzo famous.

Ferruccio was born in House 22 in Renazzo di Cento, to parents Antonio and Evelina, who were grape farmers in the area. The grapes and the resultant frizzante (slightly sparkling) lambrusco wine provided the family with a comfortable living, and Ferruccio's early years were literally grounded in the land and the grape farming lifestyle.

However, what really fascinated him from childhood was machinery, and while still a boy he set up a small working machine shop in one of the farm barns. Within this, he made and modified farm machinery, although at least once he set fire to the workshop.

Ferruccio was lucky, in that his parents understood and accepted his love of machinery, and were able to support him through his studies at the Fratelli Taddia technical institute. There is uncertainty as to whether Lamborghini studied engineering or industrial design, as well as whether he obtained his degree, or if military service obligations prevented him from completing the course.

In more than one way the Second World War was the formation of Ferruccio Lamborghini. Drafted into the Regia Aeronautica (Italian Air Force) in 1930, he was sent to the island of Rhodes, where he started off as a mechanic in the vehicle maintenance section.

Aside from his evident mechanical talent in maintaining the vehicles, Ferruccio brought himself to the notice of the base's commanding officer, and fostered a reputation by working on, and modifying, the commander's beloved Alfa Romeo. His reputation even survived after the Alfa's brake system failed following a 'modification and improvement' to the system, which ended with the car plunging into the Mediterranean. He was happy to work on just about any vehicle on the base, and was said to have burnt all the truck service manuals (having memorised them first), so that only he knew how to service and repair the trucks. Not entirely surprisingly, Ferruccio soon became the supervisor of the vehicle maintenance unit. The island of Rhodes fell to the British in 1944, and Lamborghini was taken as a prisoner of war. He then continued to work on the island, maintaining and repairing British army vehicles, until he returned to Renazzo in 1946.

Grape farming had never been his passion, and soon after returning home he identified a latent but pressing need for tractors amongst the local farmers. Italy's industrial base had been devastated by the war, and, with resources scarce, Ferruccio's first tractors were built from scrap material.

In 1947 Ferruccio married Clelia Monti, and it was during their honeymoon that Ferruccio got his big break. He happened to find out the British military were scrapping some light armoured vehicles, and with this the honeymoon was brought to a sudden halt while Ferruccio negotiated the purchase and transport of these vehicles back to Renazzo.

He stripped off the armoured plating, lightly modified the mechanicals, and thereby created a 'carioche trattori,' or 'agricole carioche' – a light tractor.

In keeping with his cherished birth sign, Ferruccio was a ferocious worker. He had a keen belief in the importance of hard work for the greater well-being of a person and society, and said "When you stop working, you start to die."

His carioche sold well, and in 1949 he opened a purpose-built factory in Pieve di Cento, not far from Renazzo, to continue with this work on a larger scale. By 1952 he was building all-new tractors to his own design, and in 1954 he started building air-cooled diesel tractors.

Soon thereafter, Lamborghini Trattori – as his company was known – became one of Italy's largest tractor manufacturers, and part of this success can be attributed to the excellent after-sales service that Ferruccio insisted his dealers provide.

Another reason given for the success of his tractors was that a relatively large percentage of his tractor components were made in-house, and with tight quality control, thereby earning Lamborghini a reputation for reliability.

In 1960 Ferruccio was part of an Italian trade delegation to the United States, and, inspired by what he had seen, he diversified into making domestic and industrial air-conditioning and heating equipment under the name Lamborghini Bruciatori. Again, part of the success of this section of his industrial empire was down to excellent after-sales care, which included the novel option of a service plan.

It is likely that it was the lack of this same after-sales care – with regards to his own expensive and exotic cars – that might have been one of the many factors that led Ferruccio to start building his own high-performance cars within the following few years, with quality and service as cardinal selling points.

The Bruciatori factory was also in Pieve di Cento, an economically deprived area at that time. By setting up a major new industry in this challenged area, Ferruccio not only received a grant from the Italian government, but also came to the notice of local and national politicians. His tractor factory, which produced one tractor a day in 1949, had increased production to 1500 a year in 1958, and increased further to 5000 a year by 1968. Lamborghini Bruciatori soon achieved the same degree of success, and by his early 40s Ferruccio Lamborghini was a very rich man, and was recognized as a top industrialist. By the late 1960s he was among Italy's wealthiest citizens, his companies employing approximately 4500 people.

With all this came honorific titles. He was awarded the Commendatore Ordine al merito della Repubblica Italiana (Order of Merit of the Italian Republic), in recognition of his contribution

Right: Ferruccio Lamborghini (centre) with Giampaolo Dallara (left).and Giotto Bizzarrini.

Above left: Period photograph of key Lamborghini personnel awaiting a flight. Left to right: Romano Artioli (of Bugatti, not Lamborghini, fame); Jean-Marc Borel (famed Lamborghini author); Ferruccio Lamborghini (red tie); Paolo Stanzani (grey tie); Marcello Gandini (far right, with yellow tie).

to Italy's industrial progress. In 1969 he received further honours, when the Italian president, Giuseppe Saragat, conferred upon him the title of Cavaliere del Lavaro (Knight of Labour in Industry).

It would not be outlandish to say that Enzo Ferrari's personal motto could have been 'divide et imperare,' or 'divide and rule.' In contrast, when answering the question "What sort of man are you?" Ferruccio Lamborghini replied "A normal person, a man who likes creating things. A good worker in the morning, and a man who likes enjoying himself in the afternoon. Because I'm not interested in ending up like my colleagues, with heart problems." He was known to be very approachable, and in the early years – although already rich and successful – could often be found with a spanner in his hand, working alongside a junior employee in his tractor or air-conditioning factory. Even after his semi-forced retirement to his vineyard, La Fiorita, near Lake Trasimeno, period film clips show him actively involved in the gritty end of grape farming, including driving a Lamborghini tractor between vine groves. One clip, filmed towards the end of the

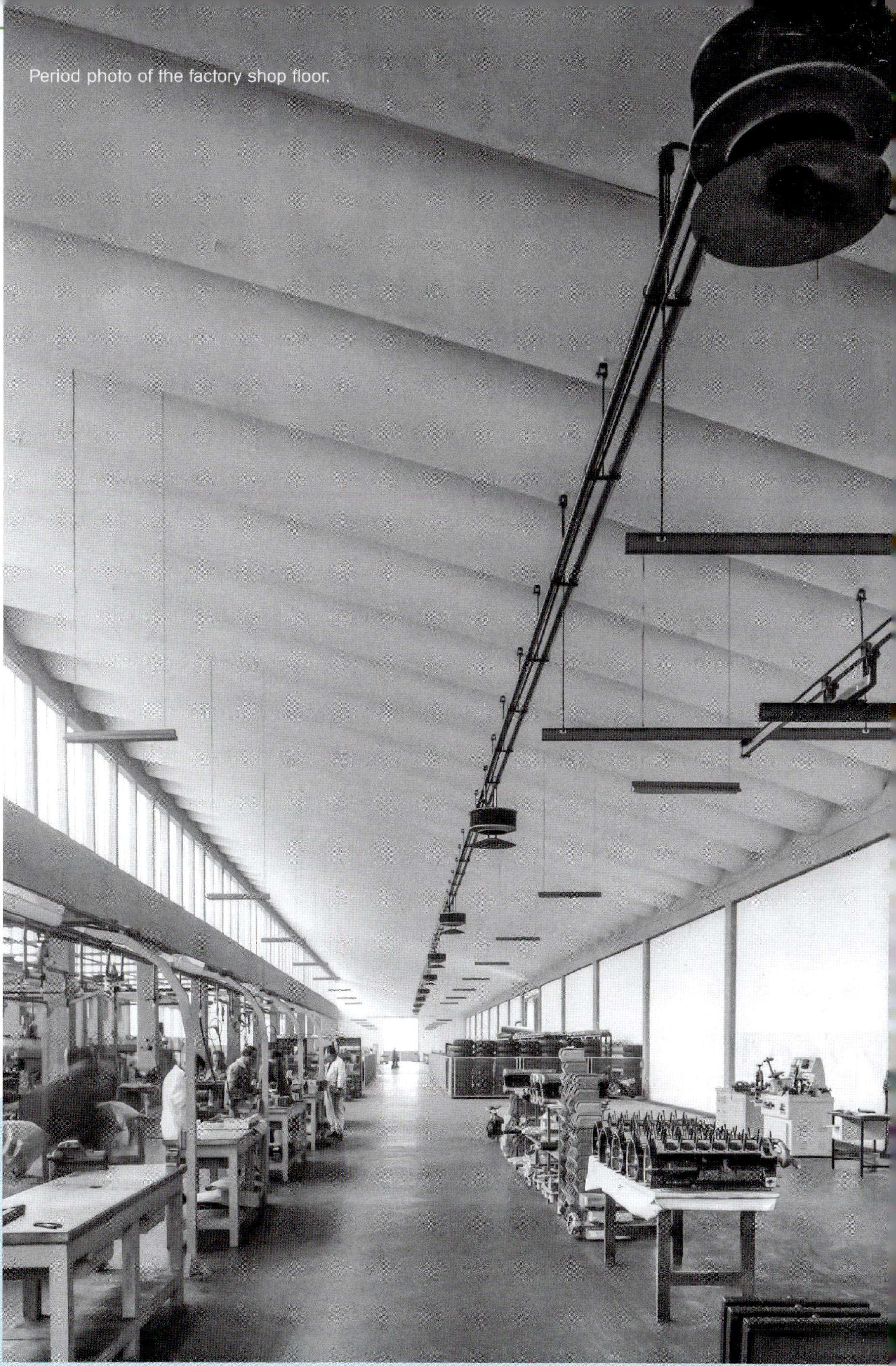
Period photo of the factory shop floor.

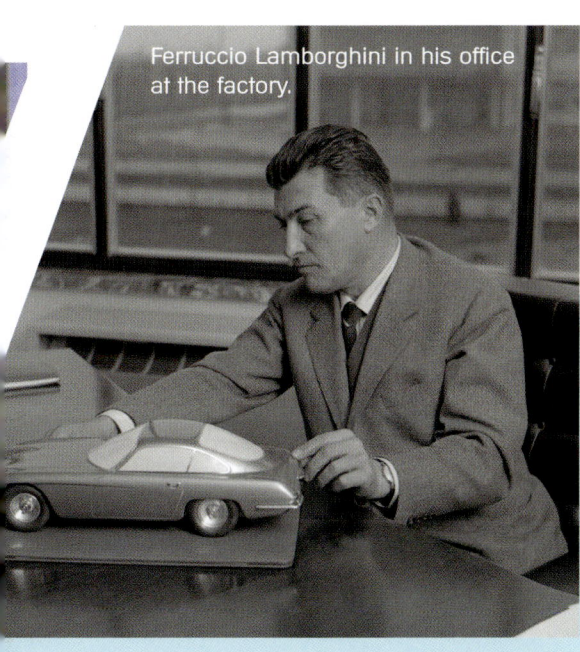

Ferruccio Lamborghini in his office at the factory.

working day, shows him driving his tractor into the gloom of a dark barn, before bursting out of it a short while later in a gleaming white Countach, and oversteering his way towards the gates of the vineyard.

Ferruccio was a jovial man, full of excitement about life, his factories and their products, good food, good wine, and desirable women. He married three times, his first wife Clelia Monti sadly passing away while giving birth to their son Antonio in 1947. The relationship with his second wife, Annita Borgatti, soon ended in divorce, and his third marriage was to Theresa Cane, mother to his daughter Patrizia.

There is a lovely story of engineer Paolo Stanzani, one of the fathers of the Lamborghini Countach, being told by Ferruccio Lamborghini's personal assistant Marisa, immediately before a long distance business trip, that Ferruccio had said "Remember one thing, we should travel on two separate planes. That way if one plane crashes, then the company will not be finished." Accordingly, Stanzani said to Ferruccio "You will be flying with Dallara, and on the following day, I will fly out with Marisa." With not a moment's hesitation, Ferruccio laughingly shot back "Ingegnere [engineer], if you don't mind, *I* will fly out with Marisa, and *you* will fly out later with Dallara." It was some time later that Dallara informed young Stanzani that Marisa was a 'close friend' of Ferruccio.

In the same vein of enjoying life to its fullest, Ferruccio's inherent love of machines and fast cars found expression in the ownership of a variety of expensive and exotic sports cars. With his fortune already made by the early 1950s, Ferruccio indulged in an Alfa Romeo 1900 Sprint, followed by a Super Sprint and then a Lancia Aurelia B20, before moving on to a Jaguar E-type, an OSCA, a Morgan, a Corvette, an Aston Martin, a Mercedes 300 Gullwing, and two Alfieri-designed Maserati 3500 GTs.

In direct contrast to his legendary rivalry with Enzo Ferrari, Ferruccio was particularly fond of, and related well to, the Italian industrialist Adolfo Orsi, the owner of Maserati. Although half a generation apart in age – Orsi being the elder – there were many parallels between Lamborghini and Orsi: their modest family origins; their early excursions into industry, and particularly farming equipment; the initially debilitating, and later devastating effects of unions and strike actions on their car-building empires; and, finally, large South American industrial orders that were never paid for, directly leading to the loss of their supercar companies. Ferruccio said of Orsi, "The owner of Maserati was a man I had a lot of respect for; he started life as a poor boy, like myself. But I did not like his cars much. They felt heavy, and they did not really go very fast."

Ferruccio appreciated the power of his Alfa Romeos, the civility and sophistication of his numerous Aurelia B20s, and the aesthetics of his early Jaguar E-type Coupé. In fact he was so taken up by the appearance of his E-type that he said "But it looked so good. When I had my first car built by Scaglione, I told him that I wanted an Italian version of the E-type." The dynamics of the E-type, however, did not find much favour with Lamborghini; "On the road I found the rear end was rather nervous, even though, on paper, the rear suspension looked great." Lamborghini was more impressed by the technical sophistication and driving style of his Gullwing Mercedes 300 SL – ironically a car whose chrome-molybdenum spaceframe chassis with rear swing axles was notorious for pronounced and unpredictable snap oversteer. He said of it: "In 1954 or 1955 I got a Mercedes 300 SL, the one with the gullwing doors. It was a remarkable car, a very progressive design for its day. No, I did not keep it. After two or three years I sold it to a friend. I had to try something new."

It was therefore almost inevitable that Ferruccio would, at some

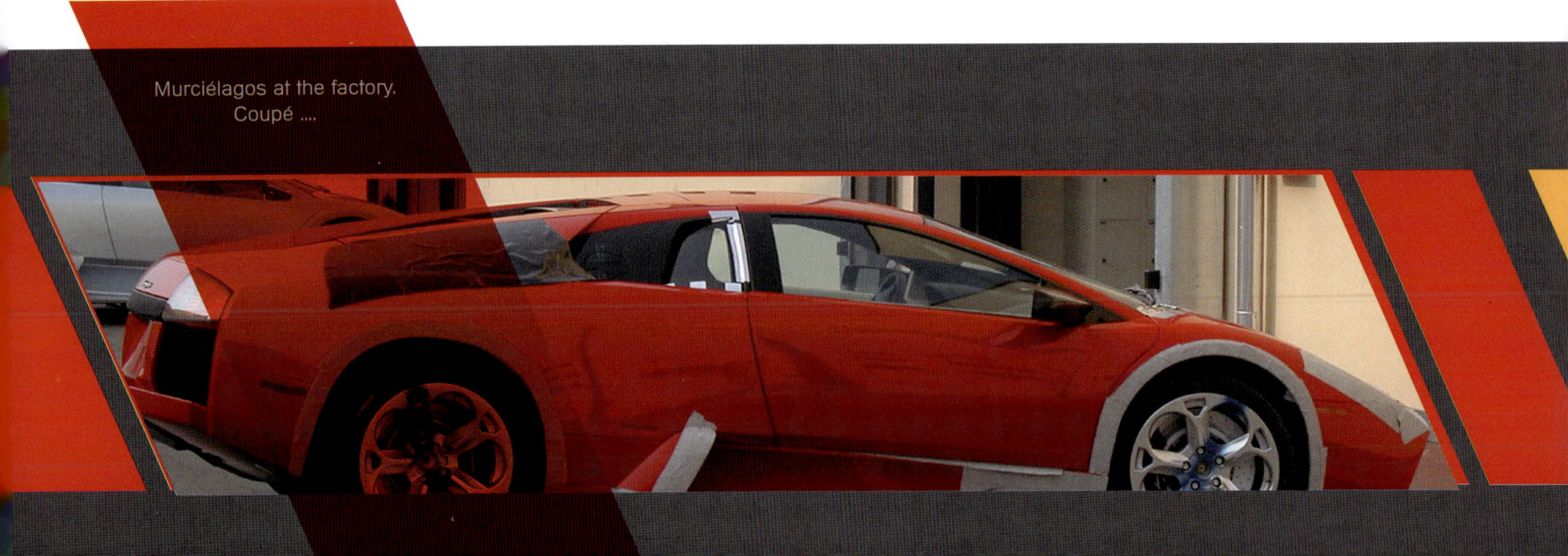

Murciélagos at the factory.
Coupé ....

stage, acquire an example of what was already, in the latter half of the 20th century, Maranello's most exotic and expensive product, and what is today, according to Brand Finance, the world's most popular and most instantly recognisable global brand. Ferrari, with its on-track activities and its stunning road cars, has displaced Coca-Cola, Disney, Rolex and Apple – amongst others – to acquire this distinction.

Lamborghini takes up the tale: "In 1958 I went to Maranello for the first time to buy a 250GT Coupé, the two-seater by Pininfarina. After that I had one, maybe two 250 GT Berlinettas, the shortbase car from Scaglietti. I did like that one very much. It was ahead of its time, had a perfect balance, and a strong engine. Finally, I bought a 250 GT 2+2, which was a four-seater by Pininfarina. That engine was very strong, and it went very well."

Lamborghini, in contrast to some reports, was generally very pleased with his Ferrari cars; "I had three or four of them. The Ferrari was a very good car, I must admit. The best I had had so far, apart from the Mercedes 300SL. After I got my Ferrari, my six other cars: the Alfa Romeo, Lancia, Mercedes, Maserati [presumably two, if the numbers are to add up], and Jaguar were always left in the garage."

But there was one recurrent issue that bothered Ferruccio; the clutches on his Ferraris. "All my Ferraris had clutch problems. When you drove normally, everything was fine. But when you were going hard, the clutch would slip under acceleration. I went to Maranello regularly to have a clutch rebuilt or renewed, and every time the car was taken away for several hours, and I was not allowed to watch them repairing it. The problem with the clutch was never cured."

According to his nephew Fabio Lamborghini, Ferruccio, together with his own tractor mechanics, stripped one of the broken down cars, and found that the clutch Ferrari was using was too small in diameter to cope with the torque output of the engine. He replaced the broken clutch with a larger diameter clutch from one of his tractors, and this absolutely sorted out the problem.

In his typically open and straightforward manner, Ferruccio thought it worth both their time to explain to Enzo Ferrari that he had diagnosed the cause of the recurrent clutch problem, and that he had also come up with a solution to the problem. Ferruccio's son, Tonino Lamborghini, explained that this was offered politely and in good faith to Enzo Ferrari.

Enzo Ferrari had at his disposal the power to grant or withhold an audience with himself, and was known to use this power mercilessly, keeping even invited guests waiting for hours. This was not something that the inherently down-to-earth, plain-speaking Ferruccio could relate to: "Every time I went to Modena, everyone seemed to take a malicious pleasure in making me hang around waiting. Ferrari's answer to my complaint on this score was that one day he kept the King of Belgium waiting, so Mr Lamborghini, builder of tractors and boilers, really had no cause to object."

Ferruccio again: "So I decided to talk to Enzo Ferrari. I had to wait for him a very long time. 'Ferrari, your cars are rubbish,' I complained. Il Commendatore was furious; 'Lamborghini, you may be able to drive a tractor, but you will never be able to handle a Ferrari properly.' This was the point when I finally decided to make a perfect car."

Ferruccio Lamborghini was a shrewd businessman, and it is unlikely that this single encounter with Enzo Ferrari would have been the sole reason for him setting up a new car company. We know that he had a fascination with mechanical objects right from childhood, and had experienced the fruits of success in two diverse engineering arenas. (Not all his early engineering enterprises were successful, however:

...and Roadster.

"After the Second World War, I had a small facility for manufacturing exercise equipment for Italian beauties who yearned to keep their figures. Business wasn't good, however, because at that time, Italian beauties were yearning chiefly for enough food to fill their stomachs.") With this in mind, Ferruccio probably saw a successful supercar company as a publicity vehicle for his other businesses, and had the penchant for recognition; and what better way to gain recognition in car-mad Italy, than to build cars that rivalled – and even excelled over – the best available?

This vigorous enthusiasm for establishing a supercar company to rival Ferrari was tempered by the risk involved with spending a vast fortune on a start-up that might fail. However, Ferruccio had a few

The famous yellow lettering on top of the factory roof.

Sant'Agata Bolognese, and in doing so established this tiny commune as one of the holy sites in the motoring world. Located in the province of Bologna, in the Emilia-Romagna region of Northern Italy, this site was close to Cento, Ferruccio's birthplace, and to his tractor, air-conditioning and heater factories. It was also close to Modena, one of the seats of the Italian motor industry, with a long and proud tradition of manufacturing and tooling. The people of this area were known to be hard-working, and many had the specialist skills and knowledge that Lamborghini would need to build a world-beating high-performance super-sports car. The municipal council of Sant'Agata was also happy to offer concessions and grants to attract new industry and capital to its area, and the entrepreneurial Ferruccio Lamborghini made full use of this.

In 1962, Ferruccio bought 90,000 square metres of land on the edge of Sant'Agata, positioned on the main road between San Giovanni and Nonantola. Specialist workers could be attracted from nearby Modena, and the good transport links also meant that raw materials could easily be brought to the factory, and finished supercars readily shipped out. By the middle of 1963, Ferruccio had already spent 500 million lire constructing the first phase of the factory, and greater expense could easily be seen looming on the horizon.

Aware that a fully-built factory was not needed for research and development work to commence, Ferruccio literally curtained off a section of his tractor factory, and the first work on a Lamborghini supercar actually started there. Meanwhile, the construction of the new factory at Sant'Agata went ahead at full speed. Lamborghini was determined that his new car factory would be state of the art in every way, and he took time off from his other business activities to personally supervise the construction of this factory. The end result comprised two long buildings, housing over 500,000 square feet of floor space, with offices on the first floor and a well-lit twin-track ground floor production line. The buildings were well thought out, with a spares department and a service department adjacent to the production lines. A service road completely encircled the factory. There was a very modern feel to the factory complex, which had a vast glass frontage, and the now-iconic word 'Lamborghini' stood out high above the first floor, in bold relief and painted in bright yellow.

advantages which he put to good use: he had prior knowledge of setting up an engineering enterprise, sound business acumen, and wasn't encumbered by having to use existing buildings or workers. This last point meant he was free to locate the facility wherever he chose, and able to seek out and make use of government subsidies.

Ferruccio elected to set up his new car company on the outskirts of

The author with Valentino Balboni at the factory museum, 2014.

Above: The 1963 Franco Scaglione conceived 350 GTV prototype, which, by 1964, became the first production Lamborghini supercar – as below: the 350GT. (Tomini Classics)

The basic structure of the factory was completed within eight months. When Ferruccio Lamborghini invited the world's – understandably sceptical – motoring journalists to view his first car, the 350 GTV, on 20th October 1963, the launch took place in the new factory. Ferruccio could not only be proud of his new state-of-the-art car, but also of his similarly state-of-the-art new factory. Fascinating cars were to emerge from within over the subsequent years, including the 350 GT, 400 GT, Miura, Islero, Espada, Countach, Urraco, Silhouette, Jalpa, LM002, Diablo, and, of course, the wonderful Murciélago. More recently, the Gallardo, Aventador, and Huracán have been the result of the alchemy performed within this temple.

In many ways, the acclaim with which the Miura was received in 1966 was the highlight of Ferruccio's tenure at the helm of Automobili Lamborghini SpA. The subsequent three decades were a turbulent time for the fledgling company, with financial hardship, recession, petroleum scarcity, and union-led challenges leading to multiple changes of ownership, and even bankruptcy.

The history of Lamborghini during this period has been extensively documented elsewhere, but the key points are as follows. The winter of 1967 saw extensive student unrest in northern Italy's universities by those unhappy with high tuition fees, poor and inconsistent lectures, and the high cost of living. There had already been massive worker agitation in the northern factories, notably the 1962 strikes at Lancia and Fiat, when about 70,000 workers walked out. It was almost inevitable that these two groups would find common purpose, and, when this happened, the scene was set for disruption on a national scale. By the summer of 1968, a new class of radical 'worker-student' groups began to supplant existing, moderate, shop-floor unions. Strike action continued throughout the early 1970s, with six million going on strike in 1973, and occupation of the Fiat Mirafiori plant taking place in 1975.

Although Ferruccio had always been considered a sympathetic, considerate, hands-on owner who enjoyed good relations with his work-force, Automobili Lamborghini SpA was not immune to the chaos sweeping through Italy. These union-led problems were compounded by the 1973-1974 stock market crash – one of the worst in modern history – during which the New York Exchange's Dow Jones Industrial Average lost 45 per cent of its value, and the London Stock Exchange's FT 30 lost 73 per cent of its value, thus severely affecting potential supercar buyers worldwide. A further major blow was the 1973 oil crisis, which saw a 400 per cent increase in the price of oil, when the Organization of Petroleum Exporting Countries imposed an embargo on those countries that had supported Israel in the Yom Kippur War. This embargo was only lifted in March 1974 after Henry Kissinger, the American Secretary of State, negotiated a partial Israeli withdrawal from the Sinai Peninsula, and set up talks on the status of the Golan Heights. However, this event also inflicted further damage on an already fragile supercar market.

The killer blow to Ferruccio's continued ownership of his car factory came in 1972, when the Bolivian government cancelled a previously placed large order for tractors. To keep his companies afloat, Ferruccio was forced into selling 51 per cent of his supercar company almost immediately, to a Swiss businessman by the name of Georges-Henri Rossetti. Within a year, he had to sell his remaining 49 per cent stake to Rene Leimer, a business associate of Rossetti. He was also forced to relinquish his tractor business to the SAME group, a rival Italian tractor manufacturer, and with this he retired to La Fiorita, a 32-hectare estate near Lake Trasimeno in Umbria, where he planted Sangiovese, Gamay, Ciliegiolo and Merlot vines. His well-regarded red wine was sold under the name Colli di Trasimeno, but it soon became known as Sangue Di Miura, or Miura's Blood. Ferruccio Lamborghini passed away on 20th February 1993, having suffered a heart attack, and is buried in Renazzo. One of his last acts as the sole owner of Automobili Lamborghini SpA was to sanction the production of the Lamborghini Countach, an act that ultimately secured the company's existence, and cemented its reputation.

In contrast to Ferruccio, Rossetti and Leimer were very hands-off owners, rarely ever visiting Sant'Agata. The clamour for the Countach kept the order books full, but the Urraco, Jarama and Espada never fulfilled their sales potential, and soon Lamborghini became synonymous with the Countach, and vice-versa. This, together with poor management (Stanzani and Wallace had left in 1975) and cash-flow problems, drove Lamborghini to the wall. In August 1978, after Rossetti and Leimer had essentially given up on the company, Judge Mirone placed the control of the company under the care of an astute accountant, Alessandro Artese. Artese appointed the celebrated ex-Maserati engineer Giulio Alfieri as Director of Engineering, and both worked tirelessly to keep the company afloat. Even so, on 28th February 1980, Automobili Lamborghini SpA was declared bankrupt, and placed into receivership.

It was at this stage that Patrick Mimran – a 24-year-old French Lamborghini enthusiast, and heir to a vast fortune based on sugarcane, flour, animal feed, plastics, banking, and shipping – came onto the scene, together with his elder brother Jean-Claude. They took over the management of the company in July 1980, renamed it 'Nuova Automobili Ferruccio Lamborghini SpA,' and promoted Alfieri to General Manager. In May 1981, the brothers bought the company for about $3 million, and Patrick Mimran gradually brought the company back to profit, and re-established its reputation.

In 1987, Patrick Mimran sold Lamborghini to the Chrysler Corporation for a healthy $25 million. The chairman of Chrysler, Lee Iacocca, was of Italian heritage, and his parents were first generation immigrants who came from San Marco dei Cavoti. Iacocca had been interested in buying an Italian supercar manufacturer for sometime, and Chrysler had already acquired a 15 per cent share in Maserati. However, the then-current Countach 5000 QV brought with it a much higher profile than any contemporary Maserati model, and Lamborghini was a much more exclusive prize for Chrysler to net. It is rumoured that the later Countach QV and Anniversary cars were better built because Chrysler was able to provide financial stability to the company and its workers, and also because Chrysler insisted on stricter quality control. The Diablo flagship was launched in 1990, under Chrysler's stewardship. While the Diablo was initially a huge sales success, and took Lamborghini to profit, by 1992 Lamborghini was again making a loss as the high purchase price of the Diablo started having a negative impact on sales. Chrysler soon wanted to off-load Lamborghini, and it found its saviour in MegaTech.

MegaTech was an Indonesia-based company, headed by Setiawan Djody and Tommy Suharto. The latter was the youngest son of the second President of Indonesia, Muhammad Suharto, who had wrested power from President Sukarno in 1967, and who remained in power until 1998. As Army Chief of Staff from 1965 to 1967, Muhammad Suharto had led an anti-communist purge that the CIA had described as "one of the worst mass murders of the 20th century." Djody already owned a stake in American supercar manufacturer Vector Motors, and hoped that collaboration between the two companies would lead to profitability. In 1995, Lamborghini was restructured, with the Malaysian company MyCom Bhd acquiring 40 per cent of the company, and an Indonesian company called V'Power Corporation, which was solely owned by Tommy Suharto, controlling the remaining 60 per cent.

The Asian financial crisis of 1997 had a direct impact on Lamborghini. Starting with the collapse of the Thai baht, most of Southeast Asia soon suffered devalued currencies, slumping stock markets and a marked rise in private debt. The devaluation of the Indonesian rupiah led to widespread rioting, and President Suharto was forced to step down. Malaysia was also hit hard, albeit not as badly as Indonesia. Ironically, 1997 was the year that Lamborghini moved into profitability, but nevertheless South-East Asian politico-economic events dictated yet another change in ownership for Lamborghini.

In September 1998, the Volkswagen Group acquired Lamborghini through its subsidiary, Audi AG, for about $110 million. The chairman of the Volkswagen Group at that time was the formidable Ferdinand Piech, a grandson of Ferdinand Porsche, the designer of the Volkswagen Beetle, and founder of his eponymous sports car company. Ferdinand Piech made his mark independently as a serious engineer, through his work on the Porsche 906 and the Porsche 917 racers, the latter giving Porsche its first overall win at Le Mans in 1970. Piech was just as successful a businessman as an engineer, and saved Volkswagen from bankruptcy in 1993, and aggressively moved into more upmarket and profitable markets, acquiring Bugatti, Bentley and Lamborghini during his tenure as chairman. At the time of the Lamborghini purchase, Audi stated that Lamborghini could strengthen Audi's sporting profile, while Lamborghini could benefit from Audi's technical expertise. It was with Audi money, under Audi stewardship, and through the pen of Luc Donckerwolke, an ex-Audi designer, that the Murciélago was launched in 2001: the first new Lamborghini in more than a decade.

Volkswagen and Audi continue to support and mentor Lamborghini to date, and Lamborghini continues to act as an advanced test bed for the Volkswagen Group, especially in the development and use of carbon composites for cars.

Ferruccio Lamborghini

THE MIURA

The oft-used term 'supercar' carries no universally accepted definition, but there are those who would argue that the title of the world's first supercar must lie with either the curvaceous, sensuous and lithe Lamborghini Miura, or its successor, the angular, aggressive and wedge-shaped Lamborghini Countach.

At its debut in the huge Palexpo halls of the 1966 Geneva Motor Show, the Miura was introduced as the fastest production road car in existence. This, together with its stunning appearance, and particularly its revolutionary engine-transmission-chassis layout, guaranteed it top billing at the show. Featuring a Bizzarrini-conceived V12 engine in a transverse mid-engined layout, there was just no other car like it.

The Miura was not the first production two-seater road car to feature a rear mid-engined layout; that honour fell to Rene Bonnet's 1962 Bonnet/Matra Djet. The Djet was, however, on a much lower rung compared to the new Lamborghini offering, with its 1108cc (later 1255cc) single overhead camshaft in-line four-cylinder engine. Equally, the Ford GT40 of the mid 1960s was not designed as a road car, but rather as a racing car. Developed together with Lola – the British racing car and engineering specialist founded by Eric Broadley in 1958, and based at that time in Slough – the Ford GT40 was conceived and built only to avenge what Henry Ford II perceived as a betrayal by Enzo Ferrari. Ferrari had been dominating long-distance GT racing in the late 1950s and early 1960s, but was suffering from financial difficulties. Ford saw acquisition of Ferrari as a short-cut to long-distance racing

success, but, after protracted negotiations, Enzo reneged on an almost-agreed deal at the very last moment. In response, Henry Ford II ordered that Ford develop a car that would beat Ferrari at Le Mans; and so a racing legend was born, in the form of the Ford GT40.

The unclothed Miura layout was first shown at the Turin Motor Show in November 1965, and drew gasps of astonishment for its audacious and compact layout, but it was the young Marcello Gandini's tightly-wrapped clothing of this chassis that finalised the Miura's ascent to the pinnacle of automotive fame.

The Miura was something that Ferruccio Lamborghini never intended to build; his vision for Lamborghini was as a purveyor of very high quality, reliable, comfortable, exclusive grand touring cars. However, the brilliant team that he had assembled at Sant'Agata – essentially comprising of Giampaolo Dallara, Paolo Stanzani and Bob Wallace, all then in their mid-20s – had dreams of racing, and worked outside company time to design the Miura in the hope that it could later be converted for racing. When the finished concept was finally presented to Ferruccio, he saw a marketing opportunity in the Miura, and consented to its build. Ferruccio's initial plan to build just a handful of these cars as an advertisement for his company, but once the prototype was shown to the public, demand for this hyper-exotic car quickly exceeded these plans.

Three regular production Miura variants were built: the P400, the P400S and the P400SV. The first customer car was delivered to the Milanese concessionaire, Lambocar, on 29th December 1966.

The P400 (standing for Posteriore 4 litri) had a 3929cc transversely-mounted rear mid-engine layout, with the V12 engine block and the gearbox cast as a single unit to save space, meaning that the engine and the gearbox shared the same lubricant. The P400 produced 350bhp at 7000rpm, and 262lb-ft torque at 5000rpm. This enabled Lamborghini to claim that this 1125kg car could reach 62mph from standstill in 6.7 seconds, and attain a top speed of 174mph. A total of 275 P400 units were produced between 1966 and the very end of 1968, when the first P400S made its production debut.

The 'S' in the P400S stands for 'spinto,' or tuned, and this model variant was essentially a refinement of the P400, with design modifications to tame both front-end lift at high speed, and chassis flex during high g-force cornering. The rear suspension was modified, much work was done in conjunction with Pirelli on new radial tyres, and the chassis was now constructed of heavier-gauge steel. Accompanying engine modifications included larger carburettors and engine intake manifolds, as well as redesigned combustion chambers and camshaft profiles, allowing a claimed maximum power output of 370bhp. Ventilated brake discs, electric windows and optional air-conditioning were also introduced with the S. Top speed and acceleration were not much improved over the original variant, however, as the new tyres and the 73kg increase in weight negated the 20bhp increase in maximum power output. 140 P400S units were built between 1969 and 1971.

The ultimate Miura was the P400SV (Super Veloce), more commonly referred to as the Miura SV. Many of these cars (although not the very first SVs) had a split sump, enabling the engine and gearbox to have their own individual lubrication systems. They

featured two striking bodywork modifications: the loss of the 'eyelashes' around the headlights at the front, and much wider rear wheel arches at the back. The SV had modified cam timing and different carburettors, and produced 385bhp at 7850rpm, thereby retaining Lamborghini's claim to be the producer of the fastest production car in the world. Between 1971 and 1973 150 Miura SV units were produced.

Lamborghini also produced one P400 Jota, a test mule that development engineer Bob Wallace extensively modified to conform with racing regulations; seven Miura SVJ units, which replicated some of the Jota's modifications onto a standard SV car; and one Miura Roadster that was actually built by Bertone, and later bought by the International Lead Zinc Research Organization and used to extol the potential virtues of using zinc alloys in cars.

"Look, there's a Diablo, too!" Miura on the tour preceding the Sant'Agata Murciélago launch, 2001.

The Miura SV – its distinct lack of 'eyelashes' and wider rear haunches set it apart from the P400S. Bertone showed off the beautiful lines of this Miura, finished in Arancio, at the 2006 Geneva International Motor Show. Miura as displayed at the Ferruccio Lamborghini Museum in Bologna.

Even today, the Lamborghini Miura remains one of the most well respected and sought-after cars in the world. Quite right too.

THE COUNTACH

If the Miura introduced Lamborghini as a genuine rival to the Ferrari road cars, then the Countach absolutely fixed and sealed this notion. If your mind's-eye image of a supercar demands it to be wedge-shaped, as it does for so many of us, then the Countach is the undisputed mother of all supercars. If you were a car-mad child, teenager or adult in the 1970s and 1980s, and had the wall space and pocket-money for one Athena poster, the chances were that an image of a Countach adorned your bedroom. To many, this is *the* definitive supercar.

Paolo Stanzani, Chief Engineer at Lamborghini from 1968 to 1975, fathered the Countach. Given a blank sheet design brief to produce a worthy successor to the Miura, Stanzani used his understanding of the Miura's shortcomings to imagine a chassis layout that would address these faults. The rear mid-engine layout of the Miura had proven a success, and had to be retained. A new longitudinal orientation of the massive Bizzarrini V12 engine did bring numerous advantages with it, and Stanzani was attracted by these; more efficient engine cooling and exhaust systems, easier engine and ancillary parts access for servicing, and a reduction in cabin noise could all be better achieved by choosing a longitudinal, rather than a transverse, mid-engine layout. Even more importantly, a longitudinally-orientated V12 had a different weight distribution, giving better directional stability and more forgiving handling in this high-powered road car, which was going to be driven by ordinary members of the public (wealth not being synonymous with talent or responsibility).

Stanzani's real genius was showcased in his decision to turn the longitudinally-orientated V12 through 180 degrees, and adopt a unique 'south-north' layout. This is probably the single key technical innovation of the wedge-shaped mid-engined V12 Lamborghinis, and this design feature – with later modification to suit a four-wheel drive system – has been carried through for almost five decades now.

Green envy: the first production Countach.

Countach 5000 QV 88½ (chassis 12399) Outside, inside and with doors elevated — an invitation to get in and drive!

Many longitudinally-orientated rear mid-engined cars have a conventional 'north-south' engine layout, with the engine at the front of the drivetrain assembly, the gearbox behind the engine, and the final drive behind the gearbox. The problem with this arrangement in a V12-engined car is that it results in a very long drivetrain, which either pushes the passenger compartment very far forwards, or else results in a very long car. Another problem with this arrangement is that the gearbox is a long way from the gearlever, which makes for lengthy gear control cables or rods, running all the way from the cockpit to the back of the car. This in turn results in less precise gearchanges, and more ponderous and less satisfying gearlever movements. With a 'south-north' orientation, the gearlever could descend almost directly from the driver's hand through the transmission tunnel, and directly into the gearbox. By adopting a longitudinal layout of this orientation, Stanzani solved all these issues at a stroke, and actually ended up with a car that was shorter both in wheelbase and overall length than the Miura.

The adoption of this unique layout-cum-orientation required some novel engineering. The drive from the V12 engine was now directed forwards to the gearbox, wherein a drop gear directed the drive first downwards, then backwards through a sealed lubrication chamber in the sump, to the differential. This required a new deeper sump – more ribbed for better cooling – and a new engine block casting.

The second striking innovation with the Countach was the upswinging scissor or guillotine doors, which have since become the signature sign for all the mid-engined flagship V12 Lamborghinis to date. These startling doors blend in perfectly with the spacecraft-like Marcello Gandini bodywork of the Countach. Gandini at this time was still working for Bertone, which had already presented two extreme wedge-shaped concept cars: the Alfa Romeo Carabo in 1968, and the Lancia Stratos HF Zero in 1970. The Gandini-designed Carabo (the name is derived from the Carabidae, a green/gold ground beetle that shares the same colour combination as this one-off concept car) had two features later seen in the Countach: the guillotine doors, and the

concealed headlights (although those on the Carabo were fixed lights beneath movable flaps, while those on the Countach were in movable pods). The Bertone Lancia Stratos HF Zero show car had originally been called the 'Stratoslimite,' which translates to 'only limited by the stratosphere;' an apt name. The even more extreme door-stop shape of the Zero (the story is that when Nuccio Bertone drove the 84cm-high Zero to the Lancia headquarters to meet its executives, he was able to enter the complex by driving the Zero underneath the barrier gates, before the guards could raise them) had to be tamed for a production car, but there is no doubt that the Countach was inspired to some extent by both these concept cars.

Gandini used a trapezoidal theme throughout the Countach, and remnants of this theme can still be seen in the Murciélago. Luc Donckerwolke, the designer of the Murciélago, has described his own work as having the "design language of the Miura mixed with the architecture of the Countach. The Countach has this really strong purist architecture. By mixing the two you get the icons of the Lamborghini."

The Countach's round tube spaceframe chassis is almost entirely hidden from view underneath the dramatic Gandini bodywork, but the Marchesi hand-welded spaceframe is every bit as iconic as the Bizzarrini engine, the guillotine doors, or Stanzani's south-north engine orientation in the Countach story, and fully deserves mention. Power-to-weight being the all pervasive determinant, in the acceleration-obsessed supercar world of the 1970s and 1980s, a strong but lightweight spaceframe was built by Marchesi for every Countach (except the show LP500 car) at its factory outside Modena, Italy.

I consider myself hugely privileged to have had the opportunity of meeting Signor Umberto Marchesi, and there is a tale to this meeting. I had been hoping to meet him at the factory launch of the Murciélago in 2001, but never did. In about 2013, I googled Marchesi and found that the company was still in rude health, and also found a contact telephone number. Then started a year-long saga, in which I started by asking, then pleading, and finally begging for the opportunity of meeting this legendary figure, who worked closely with Ferruccio

Snr Marchesi, pictured at the factory with the last three Countach chassis – each one an engineering masterpiece.

A rare moment – an autograph from the expert hand of Umberto Marchesi.

Lamborghini in the 1960s and 1970s, and whose handprint is to be found on every Countach.

His charming secretaries, in typically gracious Italian fashion, always patiently replied that he was now in his 80s, and that he no longer received visitors for papal audiences. Happily, I persevered, and after about six phone calls over a period of a year, they relented, but with the warning that Signor Marchesi did not speak any English, and had not received visitors for some time now. My wife and I turned up at the Marchesi factory in Via Tipografi at 4pm on the specified Friday afternoon in July 2014, to find that the factory had been partially shut down for our visit, and that Signor Marchesi had arranged for not one, but two English translators to be present. He took us all around the factory, explaining how the factory now worked, and contrasted it with how things were done in the past. Stacked high up, and almost touching the factory roof, were the last three remaining Countach spaceframe chassis, which Marchesi & C Srl now uses as exhibition pieces to promote and market the company's expertise.

Signor Marchesi said that he met up with Ferruccio Lamborghini at least once every two weeks during the late '60s and very early '70s, either with him travelling up to Sant'Agata, or Ferruccio coming down to his factory. The tubular spaceframe of the Countach was made of 40mm round-section steel tubes. He himself did the actual welding on the early Countach frames; welding circular tubes, as seen in the Countach, was technically more difficult, more time-consuming and more expensive than welding rectangular steel sections (as seen in the Diablo). At the height of Countach production, Signor Marchesi employed 14 welders for the project, and they produced an average of two Countach chassis each week. A labour- and money-intensive process indeed. At the end of the tour, he took us to his office, where he had already put aside some company literature for us, which he willingly autographed at our request. Business and craft talent wedded to modesty and kindness; a true gentleman from another era.

Finally, light needs to be cast on the artisanal skills required to transmute the dramatic Gandini design from paper on a drawing board, to metal on the road. The hand-fashioned alloy body of the Countach is yet another thing that makes this car so very special. While the roof is made of steel, and composites are used in several areas, including the inner chassis tub and the underbody tray, most of the rest of the Countach's body is made of hand-turned and hand-beaten thin gauge, aircraft-grade, aluminium alloy. These handmade alloy panels require specialist skills to work, and the Lamborghini panel beaters of that period were amongst the best in the world. The guillotine doors were particularly complex to make, and very difficult to align into the bodyshell aperture.

Lamborghini's internal designation for the Countach was Project 112, and 112 was later used as the first three numbers in every production car's vehicle identification number.

A total of 1997 Countach units were produced between 1973 and 1990. The first prototype was shown at the 1971 Geneva Motor Show, and was called the LP500 – Longitudinale Posteriore 5 litri – and was the purest and smoothest of any of the Countachs ever made. It had a periscope rear view mirror built into the roof, a recessed front bonnet to increase front axle downforce, and lacked the various cooling orifices seen in the production cars. However extensive modifications were needed to all these aspects to make the car suitable for road use. This priceless first prototype was destroyed in crash-testing at the MIRA facility in England, so that the Countach could obtain European type approval.

The first production Countach was the LP400, and it featured significant bodywork changes, principally to tackle engine cooling issues. A large rectangular air intake box now sat astride the rear wing on each side, and a NACA duct ran through each door into the rear flank. The production engine reverted to the tried-and-tested 3929cc unit, supposedly producing 375bhp. The first LP400 was delivered to a Milanese customer in April 1974, and 157 units of this variant were built between 1974 and 1978.

The LP400 S of 1978 featured radical exterior changes, with flared fibreglass wheelarch extensions, a deep chin spoiler, and an optional arrow-shaped rear wing spoiler. The LP400S was also the first car to sport the revolutionary Pirelli P7 tyres, massive at 345/35VR15, on the rears. Lamborghini only claimed a power output of 353bhp for this variant.

The LP500 S of 1982 was brought out, both to counter the Ferrari Boxer, and because the wider bodywork and tyres, as seen on the LP400 S, required a more powerful engine to compensate for the higher drag. The Bizzarrini-derived engine was therefore enlarged to 4574cc, and produced a more genuine 375bhp here.

Lamborghini presented the LP5000 QV (Quattrovalvole) at the 1985 Geneva Motor Show. A key character needs to be introduced here: Giulio Alfieri. Alfieri was one of the great Italian postwar automotive engineers, and a 25-year stalwart of Maserati during its most successful period. It is probably enough just to say that he was instrumental in both the Maserati 250F and the Maserati Birdcage (Tipo 61), to give the reader a measure of his ability and his achievements. Alfieri was appointed as chief engineer at Lamborghini in 1978, and was later promoted to plant manager. In response to the new Ferrari Testarossa and the limited edition Ferrari 288 GTO, Alfieri increased the stroke of the engine to 75mm, while keeping the bore unchanged at 85.5mm for a total swept capacity of 5167cc. At the same time he increased the compression ratio to 9.5:1, but, most importantly, he introduced new four-valve-per-cylinder cylinder heads, and downdraft Weber 44 DCNF carburettors. Lamborghini claimed that each QV produced at least 455bhp at 7000rpm and 369lb-ft at 5200rpm. Various different contemporary motoring journals independently tested the 1488kg QV, and found that it took between 4.2 and 5.1 seconds to accelerate from standstill to 60mph. The factory claimed an officially verified top speed of 183mph for the QV.

A blast from the park – still way ahead of its time today, the Countach attracts attention wherever it may go.

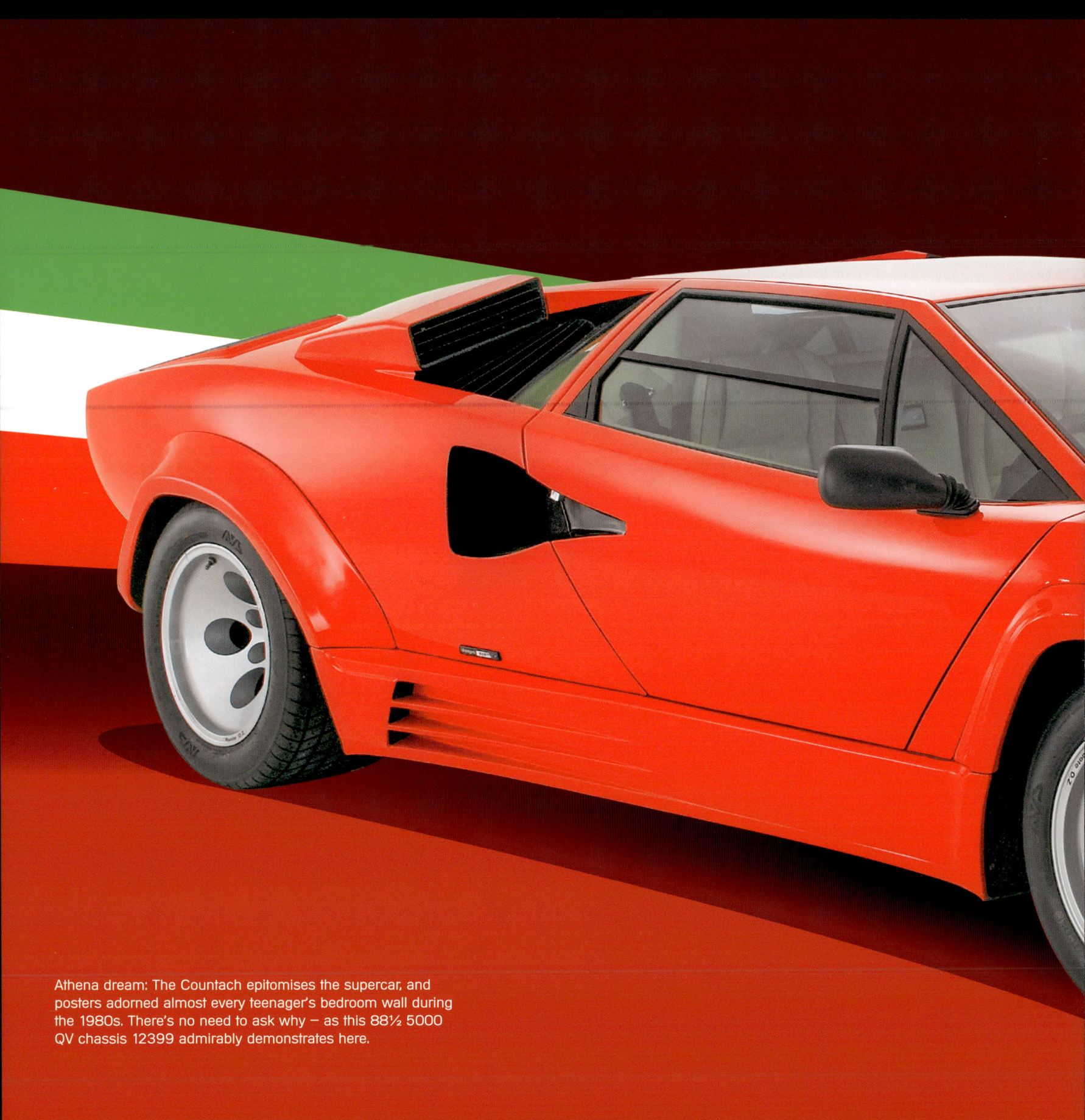

Athena dream: The Countach epitomises the supercar, and posters adorned almost every teenager's bedroom wall during the 1980s. There's no need to ask why – as this 88½ 5000 QV chassis 12399 admirably demonstrates here.

A sub-variant of the QV was the 88½ QV, which was produced between late 1987 and the introduction of the Anniversary in 1988. The 88½ QV cars had straked sills, designed by none other than Horatio Pagani of Zonda and Huayra fame, covering the dolphin belly, and these strakes fed cooling air to the rear brakes. The side strakes on the 88½ QV are different to those on the Anniversary, in that in the former the strake panels appear separate from the rear wheelarch, while on the Anniversary they appear to be in one single piece. When the author met Signor Pagani in 2001, and then again in 2014, he seemed proud of having designed these side strakes for the 88½ QV. During a lunch conversation between the author and Valentino Balboni at Sant'Agata in 2015, the veteran test driver confirmed that the 88½ QV cars had slight front and rear suspension geometry alterations compared to the earlier QVs. Valentino also said that by the time the 88½ QV cars were in production, Lamborghini had the reassurance of Chrysler's financial clout behind it, and that this meant that resources were more plentiful (or at least less scarce), and that quality control was much better than it had been previously. While North American specification 88½ QV cars had an electronic ventilation control system, many European-spec 88½ QV cars retained the traditional manual ventilation lever controls. 610 QV units were produced in total, of which something like 145 were 88½ QV cars. Just 14 right-hand drive 88½ QV cars were made.

The Countach 25th Anniversary was built from September 1988 until January 1990, and featured many interior and exterior alterations, but was mechanically very similar to the QV, save for some slight suspension geometry changes, and the adoption of new Pirelli P Zero tyres. The Anniversary had a more 'smoothed over' exterior, with reprofiled flank air-intakes, body-coloured NACA Ducts, new front and rear bumpers, new multi-piece OZ wheels, altered taillights and a new engine lid design. On the inside, there were now new two-piece seats in place of the classic banana seats, a new steering wheel hub cover, a golf club-like gearlever handle and electric windows. 657 of these units were built over a 16-month production run.

Incidentally, the famous rear wing spoiler on the later Countachs was always an option, and never a standard fitment. At times this was a very expensive option ($5000 in the mid 1980s has been mentioned). However, when chassis 12399 (carrying the F9200YR number plate) was ordered in 1987 for a 1988 delivery, the rear wing was a no-cost option. The Countach has a rear weight bias, and does not suffer from rear axle lift; if anything, it suffers from some front axle lift at high speed. The rear wing is therefore purely cosmetic, as Lamborghini's engineers zeroed out the wing's angle to make it effectively non-functional. The rear wing does, however, carry a weight and drag penalty, and is said to lower the top speed of the Countach by up to 10mph.

5000 QV 88½ Countach and friends at Wilton House's annual supercar day.

The Anniversary model was a fitting tribute to a quarter-century of Countach production, and the styling was both outrageous and apt for the time. The bonnet mascot was not included!

THE DIABLO

Perhaps unsurprisingly, the Diablo – the new Lamborghini flagship of 1990 – was not equal to the task of eliciting the same degree of shock that the Miura and the Countach had engendered at their debuts. These latter two cars introduced the transverse rear mid-engine layout, and the south-north longitudinal rear mid-engine layout, respectively. From an aesthetic standpoint, the tightly-knit bodywork of the Miura makes it amongst the most beautiful cars ever built, while the Countach stunned the world with its angularity and its guillotine doors. In fact, it is only with the futuristic Lamborghini Terzo Millennio of 2017 – an all-electric hypercar concept, designed to utilise supercapacitors rather than conventional batteries – that Lamborghini has again stunned the public in the same way that it did with the Countach, but the Diablo was nonetheless stunning in its own way. Meanwhile, the principal technological advance of the Diablo range would not surface until the introduction of the four-wheel drive system, as seen in the Diablo VT of 1993.

The Diablo (meaning 'Devil' in Spanish, and so had to be sold as the Lamborghini Costanga in Mexico, to soothe cultural sensitivities) made its debut – just outside and beneath the terrace of my wife's favourite restaurant, Le Louis XV – in the Mediterranean cold of January 1990. Yet again it was named after a fighting bull, and designed principally by Marcello Gandini. (Chrysler's Detroit designers thought that they could better the maestro's penmanship, and, thanks to the power of the dollar, they had their way in tampering with Gandini's design.)

No prizes for naming the model here – the SV was a worthy successor to the Countach.

Diablo in the detail ... The Diablo GT was a limited run, race-oriented and lighter Diablo, with carbon fibre rear wing and front air dam, shaped to accommodate a larger oil cooler. Above this on the bonnet was a large air scoop and two NACA air ducts. As well as a wider track at the front, the GT was fitted with 3-piece wheels by OZ. The interior was radically minimal to add to the atmosphere of the race-specification. Under the engine cover sat its six-litre V12 developing 575hp. Just 80 GT units were built.

Last of the line: The VT's 6.0, V12, 550hp engine powered the Diablo swansong; a fitting tribute to the success of its eleven-year production run. The final versions hint at both the heritage of Lamborghini's iconic designs, and also what was yet to come, in the shape of the Murciélago. This can be seen in the styling – from the lights and front air intakes, the one-piece wheels, akin to the Countach, and a generally cleaner, smoother body. The final run of the Diablo was the VT 6.0 SE, which was available in two colours: Orro Elios Gold, and Marrone Eklipsis, a deep maroon colour, with a bronze flip. Just 40 of this variant were produced, and command high values.

A primary design brief for the Diablo was that it had to be able to reach at least 315km/h, and, while Lamborghini claimed that it could reach 328km/h, Sandro Munari actually reached a top speed of 340km/h while testing at Nardo. The new car had a 48-valve 5707cc multi-point fuel-injection engine, which produced 492bhp at 7000rpm and 428lb-ft torque at 5200rpm. With a kerb weight of 1576kg, it was able to accelerate from standstill to 62mph in 4.09 seconds.

The Diablo VT of 1993 featured a redesigned dashboard, amongst many interior and exterior modifications, standard power-steering, and an electronic Koni-equipped suspension. The main advancement with this new variant, however, was the new all-wheel drive system – the VT standing for Viscous Traction. Through the employment of a viscous central differential, up to 25 per cent of the engine's torque and power could be directed to the car's front axle in the event of rear wheel slip. A development of this original system would later be used in the Murciélago.

A lightweight two-wheel drive street-racer called the Diablo SE30 was also released in 1993, to celebrate Lamborghini's 30th birthday. Eschewing many luxury features like power-steering, the audio system and air-conditioning, but gaining an automatic underbonnet fire extinguisher system and a 525bhp engine, the 0-62mph dash was cut to 4.00 seconds, with a claimed top speed of 333km/h (207mph).

The Diablo SV was introduced in 1995 as the entry level Diablo, and was the cheapest of the range. It had a 510bhp engine, and was rear-wheel drive with a four setting traction control system.

1995 also saw the introduction of the Diablo VT Roadster, with a much redesigned body, particularly at the rear; the detachable carbon fibre hardtop could be latched onto the rear bodywork overlying the engine lid, for the alfresco experience. 1999 saw further developments of both the VT and the VT Roadster , as well as the debut of the Diablo GT. Lamborghini's press release at the time claimed that the GT was the fastest production car in the world, with a 338km/h top speed, and the ability to accelerate from rest to 62mph in 3.90 seconds.

The Diablo VT 6.0 marked the intersection point between the Chrysler/Megatech/MyCom/V'Power ownership of Lamborghini, and the takeover by the Volkswagen Group, and being placed under the stewardship of Audi. The 6.0 update was the first task undertaken by new Audi appointee Luc Donckerwolke, on his journey towards designing the all-new Murciélago. Released in 2000, the 6.0 had a new 5992cc engine producing 550bhp, wider front and rear tracks, a new air-conditioning system, and a new carbon fibre-rich interior. The 6.0 could also be had as a special order, rear-wheel drive only car, if so desired. Many considered the 6.0, with its Audi input and accompanying quality control, to be the most desirable of the Diablo series.

THE SEVEN DEADLY SINS
& BUYING THE COUNTACH

Of the seven deadly sins that grew out of Evagrius Ponticus' original 'eight evil thoughts,' all bar one can reasonably be associated with the Murciélago.

Gula (gluttony), luxuria/fornicatio (lust/fornication), avaritia (avarice), superbia (hubris), ira (wrath), and vanagloria (vainglory); all these do not seem so distant from the excesses of a technicoloured, wedge-shaped, leather-swathed, carbon fibre-bodied, V12-powered missile for two.

Even Tristitia (sorrow and despondency) has a place here, both in the hearts of besotted enthusiasts not fortunate enough to own one, and equally in the hearts of owners, when a turn of the ignition key fails to awaken the slumbering giant of an engine, or when the dashboard warning indicators light up like an unwelcome rainbow.

Under such circumstances, said enthusiasts and owners would be well advised to seek solace in the epic poem by Prudentius (circa 410AD), the *Psychomachia*, or Battle of the Soul. The poem describes the seven contrary virtues, and those

affected should practise, in particular, patientia (patience), temperantia (self-restraint), humilitas (humility) and industria (persistent effort).

Acedia (sloth), appears to be the only cardinal vice that does not marry up with the Murciélago.

Unsurprisingly, therefore, the Murciélago was unveiled with the seven mortal sins as the theme at its debut event.

On a cold and windy early autumn evening in 2001, specially-selected journalists, Lamborghini dealers, and favoured owners were driven to the northern foothills of the Mount Etna volcano in Sicily, where, at a height of 1500 metres, a stage had been set up. Mount Etna had erupted just weeks earlier, and the volcano was still smoking. This launch setting was befitting of a car that was spectacular in its beauty, but also frightening and intimidating in some ways, and which therefore demanded that lust be tempered with caution.

The unveiling started with a 4.5-minute film (actually shot in the Andes mountains), projected onto a giant screen, with

The Murciélago brought with it not only new thinking, new customers and cooperative ownership, but a whole new attitude and influence towards car design generally. Yet again, Lamborghini had re-written the rulebook – 'si applicano nuove regole!'

A glimpse beyond those guillotine doors into one driver's uniquely customised space; light, airy and ready for business.

Sant 'Agata plays host to a worldwide audience of Lamborghini owners and enthusiasts – and a very exclusive traffic jam – at the Murciélago launch.

accompanying special lighting and sound effects. A dance troupe then played out the seven capital vices, set against artificial lava streams, thundering music and make-believe volcanic smoke.

Towards the end of the 45-minute show, a metallic black Lamborghini Murciélago drove past the audience at 90km/h, and soon after another two cars were presented to the onlookers: a yellow car and another black car.

On a personal note, I was lucky enough to be invited to the public launch of the Murciélago at the factory on 8th September 2001. This came on the back of having bought a one-previous-owner Countach 5000 QV 88½ (chassis 12399) on 26th May 2001 from HR Owen in London. At the time this was the only authorised Lamborghini dealer in the United Kingdom.

The letter of invitation from the factory, which had been informed of my existence by HR Owen, was beautifully phrased, asking that I bring my Countach back to its birthplace. At the time, my only interests were in the Countach and the hallowed Sant'Agata factory,

and to be honest the Murciélago was merely a sideshow to me. This, however, was a great excuse to go to the factory.

We were allowed onto the sacred grounds of the factory at about noon, and were pretty much free to wander around at will, except for one part of the factory complex, where the Murciélago was to make its debut. I was particularly fascinated to see for myself the famous terracotta tiled floor of the production line, and the numerous Countachs that owners had either driven down or flown in from around the world, which were parked all over the factory grounds. I was even attracted to the paper napkins in the workers' canteen, which bore the Lamborghini logo, and I collected a few. Also on display, at an alarming 45-degree angle atop a large lorry-drawn 16-wheeled trailer, was a twin-hulled Class 1 World Championship off-shore powerboat, powered by Motori Marini Lamborghini engines.

The highlight of the whole event was held that evening, in the part of the factory that had previously been cordoned off.

Round tables seating ten, topped off with crisp white tablecloths, were set before a darkened stage. The waiters were efficient in serving the dishes and in replacing the empty wine bottles, but food and drink were very much peripheral issues here.

Quite suddenly, the room lights started to dim, and with this Steppenwolf's *Born to be Wild* started blasting through the speakers. Slightly disorientated by the low light and the loud music, one was then startled by the nearby roar of what was clearly a multi-cylindered engine. The revs were allowed to drop to idle, only to be repeatedly sent up to meet the limiter, and with each such cycle, the tension and sense of anticipation in the room grew. The stage then became progressively more lit up, and you could hear and feel a car approaching the room. Embarrassingly, just as it was about to make its stage entrance, which involved driving up an incline hidden by a screen from the guests, the engine stalled and died, not

once, but twice. Nevertheless, it simply added to the drama, which climaxed when a jaw-dropping, eye-popping, knee-trembling Verde Ithica (fluorescent green) Murciélago burst onto the stage. Just about everyone rushed to see the car close-up, and it was soon mobbed by a sea of grown adults. It was a good ten minutes before I could get my 20 seconds of seat time, in what was now my second-favourite Lamborghini, and I feel very lucky and privileged to have been present at the factory launch of the Murciélago.

BUYING THE COUNTACH

The Countach had instantly become my dream car after I first set eyes on a magazine photograph of the sleek LP500 prototype, at the age of nine.

Three decades were to pass before potential ownership became a realistic proposition. When it did, I wasted little time in setting

One careful owner – the deal was struck, and this moment marked the beginning of many new adventures.

about the task with gusto, and few stones were left unturned in the search for my ideal specification Countach.

I contacted HR Owen very early on in the search, as the only authorised Lamborghini dealership in the United Kingdom at that time, and was put in touch with their sole salesman, Jason Barker. I explained what I was looking for in detail: any Countach other than a 25th Anniversary, to be in as pristine a condition as any secondhand car could be, unmolested with absolutely no post-factory modifications, and having only had one previous owner. This last request raised Jason's eyebrows, but we kept in touch intermittently while I chased up whatever other leads I could find throughout the country, and in continental Europe, all without success. I was once told that the only place that I could realistically expect to find a one-owner Countach would be in Saudi Arabia, but

> *... Of the seven deadly sins ... all bar one can reasonably be associated with the Murciélago ...*

The writing's on the wall — The 6.2 VT Coupé in profile, juxtaposed against an urban background.

DESIGN INFLUENCES

Head of Design at Centro Stile, Luc Donkerwolke readily admits to not liking open cars because of the inevitable loss of torsional stiffness, and also because a fixed roof allows for a clean sweeping arc from the front to the back of the car, a styling feature that is lost with an open-top car. Donkerwolke says that he was inspired by 3 quite disparate things when designing the open-top variant of his neo-classic Murciélago VT coupe – the B2 Stealth Bomber, the Ciutat de les Arts i les Ciencies buildings in Valencia, Spain, and the 118 WallyPower yacht.

Above left: Ciutat de les Arts i les Ciences, Valencia.

Right: B2 Stealth bomber.

Above left: WallyPower's 118 superyacht.

to expect to find plenty of sand not only in the car's interior, but also within the car's cylinder bores.

Then one day, after almost two years and quite out of the blue, I had a telephone call from Jason asking me to come up to London to see a car that supposedly met all my criteria. Work dictated that the earliest that I could get to London was five days later, on a Saturday. I met up with Jason at the famous Old School Lane workshops, and there, in a corner of the voluminous garage, sat chassis 12399; an 88½ Countach 5000 QV, Rosso Siviglia with a tan interior and red piping, one previous owner who had owned three Countachs prior to this one, and solely serviced by HR Owen throughout its 13-year life.

In those days, the Countach was nowhere as highly sought-after, and therefore not as highly priced as they are today, but it was still a very major purchase for me. With all the boxes ticked, and the price being just within the financial envelope available to me, my primary concern was that the car had already been placed in HR Owen's advertisement for the following day's *Sunday Times* motoring supplement. I was desperate to put down a holding deposit, however small, to secure the car. However, by now it was past lunchtime and, with the accounts department closed, a deposit could no longer be taken. I was genuinely worried that I would be out-bid once the car's availability became public knowledge, and to make matters worse, I was operating all day on Monday and Tuesday, and could not be contacted during working hours until Wednesday morning.

(continued on p47)

Begging for ... attention from any angle – even in a relatively subtle shade of Grigio Antares (Titanium) the Murciélago is impossible to ignore – and why would you?

Fun with some 'family members' on North American roads, the Murciélago, along with two Gallardos, looks comfortable in its surroundings – you can almost hear the symphony of engine and exhaust notes!

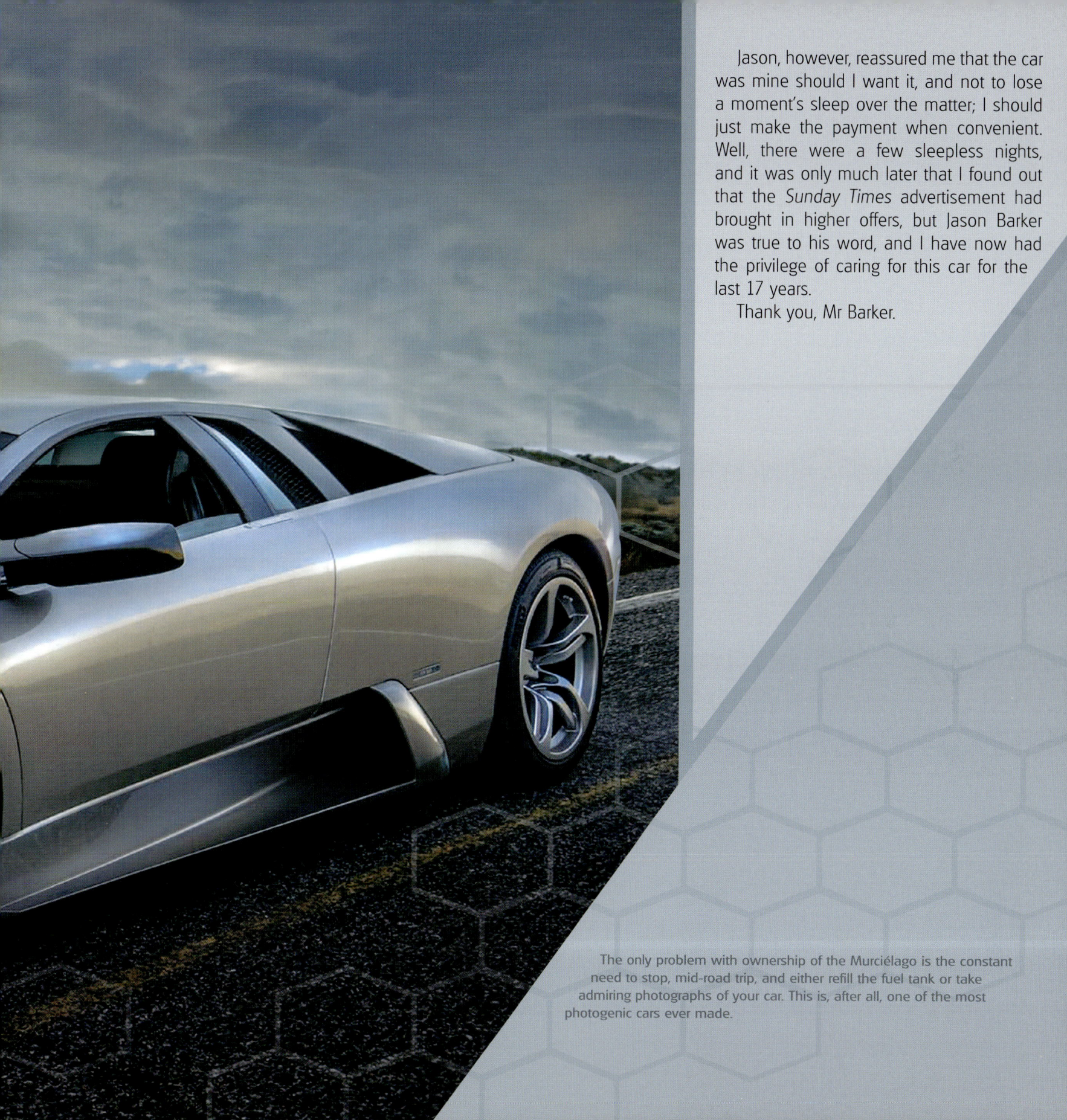

Jason, however, reassured me that the car was mine should I want it, and not to lose a moment's sleep over the matter; I should just make the payment when convenient. Well, there were a few sleepless nights, and it was only much later that I found out that the *Sunday Times* advertisement had brought in higher offers, but Jason Barker was true to his word, and I have now had the privilege of caring for this car for the last 17 years.

Thank you, Mr Barker.

The only problem with ownership of the Murciélago is the constant need to stop, mid-road trip, and either refill the fuel tank or take admiring photographs of your car. This is, after all, one of the most photogenic cars ever made.

NAMING THE MURCIÉLAGO

Born on the 28th April 1916, Ferruccio Lamborghini took his zodiac sign, the Taurus, very seriously.

The term 'zodiac' derives from the Latin zodiacus, which in turn comes from the Greek, zōidiakos kuklos, meaning 'circle of animals.' The ancient Greeks inherited their knowledge of astronomy, and their belief in astrology, from the Mesopotamians, who considered the Taurus – which they knew as the Bull of Heaven – as the first sign of their zodiac. They gave the Taurus such prominence because, at the time of their civilisation, the sun rose through the Taurus constellation on the day of the vernal equinox.

Red rag to a bull: A Bob Forstner styled Murciélago threatens the matador ...

The iconic logo was born from the very roots of Lamborghini's founder, and developed into a legendary badge that was influenced by a beast whose power, cunning and survival instincts are the key values associated with the company today – like any brilliantly conceptualised logo, it is instantly recognisable the world over – in the same way as the cars are.

The year before Automobili Lamborghini was officially established, Ferruccio Lamborghini visited a ranch renowned for breeding Spanish fighting bulls. It was this visit that inspired him to choose the raging bull symbol as the logo for his forthcoming supercar company. The ranch was originally established by Don Eduardo Miura Fernández in 1842, and is located about 60 kilometres to the east of Seville, the capital of Andalusia, in the south-west of Spain. At the time of Ferruccio's visit, the ranch was run by Eduardo and Antonio Miura, sons of the late Eduardo, and was already famous for their Miura line of fighting bulls.

The Miura breed has its origin in five historic bull breeds, namely the Navarra, the Cabrera, the Vistahermosa-Parlade, the Veragua and the Gallardo. The Miura breed is known for being large, difficult and fierce, but most of all for being cunning – so much so that matadors have the saying, "never turn your back on a Miura."

Some of us just consider bull fighting blood-thirsty and cruel, and that its unequal nature means that it is a cowardly activity, which debases the word 'sport' and reflects particularly badly

on the spectators, who sit safely out of the way of all danger. Ernest Hemingway, who famously worked as a reporter on the Spanish Civil War in the late 1930s, thought otherwise, and had become a bull-fighting aficionado after a much earlier visit to the Pamplona fiesta in the mid 1920s. He wrote the classic book *Death in the Afternoon*, which details the traditions and ceremony of bullfighting in Spain, and also delves into the nature of fear and courage. Hemingway wrote in this book, which was published in 1932, "There are certain strains of bull which have a marked ability to learn from what goes on in the arena … faster than the actual fight progresses, which makes it more difficult from one minute to the next to control them … these bulls are raised by Don Eduardo Miura's sons from old fighting stock."

Ferruccio Lamborghini named his first mid-engined supercar after this breed, and almost four decades

Fast charge: The Murciélago bellows past, yet it's only hinting at the remaining raw power that lies behind the occupants. Cunningly, like the bull it was named after, there's plenty of brute force left in reserve from the 571 hp available.

later, the Murciélago supercar was named after one particular bull, which was descended from one of the five old fighting stock breeds that had originally contributed to the Miura breed – the Navarra.

On 5th October 1879, this particular Navarra bull, going by the name Murciélago, fought against Rafael Molina Sánchez, at the Coso de los Califas bullring in Córdoba, Spain. Sánchez (1841-1900), was at the height of his powers at this point in his career. He had become a fully fledged matador in 1865, and retired in 1893, having been gored seven times. Murciélago survived 90 sword strokes, of which 24 were said to be deep stab wounds, but continued to

fight with such vigour, spirit and passion, that the spectators called out for his life to be spared. This was an honour that only the matador could bestow, and on this occasion Sánchez – who was also known as 'Lagartijo,' or lizard – acceded to the crowd's request.

Murciélago was later presented as a gift to Don Antonio Miura and Don Eduardo Miura, and they entwined Murciélago further into the Miura line, by siring him with 70 Miura cows.

Roadway to freedom: Sant'Agata's terracotta floor is an historic part of the Lamborghini factory. At the time of writing, around 32,000 hand-built cars have passed through since 1963.

THE MURCIÉLAGO
VT 6.2 COUPÉ

The Lamborghini Murciélago VT 6.2 Coupé was Lamborghini's first all-new model under the ownership of the Volkswagen Group, and was also Lamborghini's first new design since the 1990 Diablo. The Murciélago – launched in the foothills of the Mount Etna volcano in Sicily, in September 2001 – was an evolution of the Diablo, which itself was a development of that seminal supercar, the Countach. The fundamentals of the Murciélago were carried over from its two illustrious predecessors in the shape of the spaceframe chassis, the Bizzarrini-derived V12 engine, the unusual engine-gearbox orientation, and even in having genuinely vertical-opening guillotine doors. Supposed 'revolutionary' advances would have to wait until 2011, and the advent of the Aventador.

Prior to the Murciélago, prototype Lamborghinis were mainly tested on the roads around the Sant'Agata factory (one particularly notorious stretch even carries the unofficial name of 'Balboni Boulevard,' or 'Balboni Highway'), and occasionally at Nardo.

In 1971, there was only one LP500 Countach prototype produced, and, almost 20 years later, with the development of the Diablo, only five prototypes were produced.

Following Lamborghini's takeover by the Volkswagen Group, its new owner felt that its own reputation for quality and reliability not only had to be safeguarded, but that these world-leading qualities needed to become features of its newly acquired marque also. No longer would it be acceptable to have early-adopting new owners act as unpaid test and development drivers!

To this end, 12 test cars were built on production line one, before production began. Test and development drives did continue around the factory, but test cars were also shipped out to the United States for high temperature and low humidity testing in the desert, and further high speed testing now took place at the Nürburgring and Imola, as well as at Nardo.

The Murciélago continued the traditional Lamborghini supercar layout, with a rear mid-engined V12, the typical Lamborghini transmission orientation of the gearbox mounted in front of the engine, the rear differential forming part of the engine unit, and having a permanent four-wheel drive system with central viscous coupling.

This layout basically dates back to Paolo Stanzani's bold, original and ingenious idea of a south-north engine set-up, and had the advantages of placing all the major sources of mass close to the centre of the car. It did not need long linkages for gear changing, as the gearbox was immediately below the driver's hand, and was compact.

The layout called for a complex arrangement of first sending the drive forwards from the engine to the gearbox, then sending it downwards from the gearbox through a drop gear, before finally being directed backwards to the rear differential through a sealed tube in the engine sump. Back around 1969-1970, when this layout was originally mooted and development first started, there was a fear that this arrangement would be too complex for reliable use in a production car. However, by the time of the Murciélago's introduction in 2001, Lamborghini had been very successfully using this layout for almost 30 years. It afforded the Murciélago an optimal weight distribution of 48 per cent front and 52 per cent rear, with major resultant advantages in stability, traction, braking and handling.

THE V12 ENGINE

There is a special romance that comes with a V12 engine, and particularly so when one such engine – albeit with evolutionary modifications – stands the test of time, powering an iconic brand's flagship cars for 47 years.

This was the case with the Bizzarrini-designed 60-degree Lamborghini V12.

For me, the two outstanding attractions of the Countach, Diablo and Murciélago, other than their spectacular wedge shapes, are, firstly, the south-north longitudinally-orientated Bizzarrini V12 engine, and, secondly, the intricate spaceframe chassis – round tubed in the Countach, and square tubed thereafter. Both these features were lost forever with the demise of the Murciélago; a sad day.

Although designed solely by Giotto Bizzarrini (there appears to be little foundation to the usually brilliant LJK Setright's suggestion that Honda designed the veteran Lamborghini V12 engine. Bob Wallace, the legendary New Zealand-born Lamborghini test engineer of the time, in his typically direct and taciturn manner, dismissed this idea as "crap"), there were three other men who were fundamental to the engine's success and longevity: Giampaolo Dallara, Paolo Stanzani and Giulio Alfieri.

Each of these three men was a true engineering genius in his own right, directly involved in the development of groundbreaking race or road cars, and also critical in ensuring that Lamborghini Automobili survived through the turbulence of the 1970s-1990s.

But first things first, and back to the origins of the V12 engine. By the beginning of the 20th century, the in-line four-cylinder engine had gained predominance by virtue of its power output, its simplicity of cranking, its compact size (and so access for servicing parts), and because the casting and machining technology of the day allowed for reliable production of these engine blocks.

Henry Ford once said, "A car should not have any more cylinders than a cow has teats." However, the quest for more power, smoother power delivery, and also exclusivity as a marketing tool, led to the development of six- and eight-cylinder engines in the automotive sector. In parallel, aviation and marine demands also led to the development of multi-cylinder engines.

The world's first V12-powered car was the Schebler Roadster of 1908. George Schebler of Indiana, USA, had already invented a carburettor bearing his name, when, together with Philip Schmoll, he developed a V12 engine with valves in the head, and the cylinders set at 45 degrees.

The 12-cylinder engine has an inherent mechanical balance, while also being alert to the throttle pedal, and strong in its potential power delivery. Phil Hill, the 1961 Formula One World Champion, when speaking of the Ferrari V12, praised its "multiple stages of performance" and "beautiful flexibility," elaborating that "the more revs you use, the more torque and power are available."

Damon Hill, the 1996 Formula One World Champion, called the V12 pulsation pattern "the sonic equivalent of strawberry mousse and cream."

Herbert von Karajan, the Salzburg-born principal conductor of the Berlin Philharmonic for 35 years, wrote to Enzo Ferrari saying "When I hear your 12 cylinders, I hear a burst of harmony that no conductor could ever recreate."

THE LAMBORGHINI V12 ENGINE

So the scene was set for Ferruccio Lamborghini needing – not wanting, but actually *needing* – at least a 12-cylinder engine, with which to take on and hopefully conquer the established opposition in his quest for a 'perfect car.'

When Lamborghini mentioned his need for such an engine in his discussions with the Modena-based coachbuilder Neri & Bonacini, it recommended Giotto Bizzarrini. Bizzarrini was born into a wealthy Livornese family in 1926, and followed in the family tradition by going to the University of Pisa, where he studied mechanical engineering. He graduated with the highest marks, having redesigned a used Fiat Topolino engine for more power and better weight distribution (a rarely-mentioned parallel with Ferruccio Lamborghini's own modification of a Topolino engine for his Mille Miglia attempt), and, on the strength of this, got a job at Alfa Romeo. He was unhappy almost immediately upon arrival, after being assigned to the department responsible for the chassis rather than the engine. His time at Alfa was, however, not entirely unfruitful, in that he met and befriended Carlo Chiti there, and soon after they both moved to Ferrari. Bizzarrini became Chief of Racing Car Development at Ferrari, while Chiti became Ferrari Chief Engineer, and these two men worked on, and were responsible for, some of the most special Ferrari engines and cars of all time, including the Testa Rossa V12, the 250 SWB and the sensuous 250 GTO.

Both Bizzarrini and Chiti were caught up in the Great Ferrari Walkout of 1961, and together they set up Automobili Turismo e Sport (ATS), with Bizzarrini later founding Societa Autostar, an engineering firm through which he would handle freelance projects.

It was through Societa Autostar that, in 1962, Bizzarrini received the commission to design, and then supervise the assembly of, a V12 engine for Lamborghini. The fee was 4.5 million lire, half of which was paid on commencement of the design work, and the second half of which was to be paid at the end of the project. The contract stipulated an engine output of 350bhp, with incremental bonus payments and penalty deductions for over- or under-achieving that target, respectively.

Giotto chose a cylinder bore of 77mm and a stroke of 62mm to give a total swept capacity of 3465cc. He designed the engine to have twin overhead camshafts per bank, dry sump lubrication, and 38mm

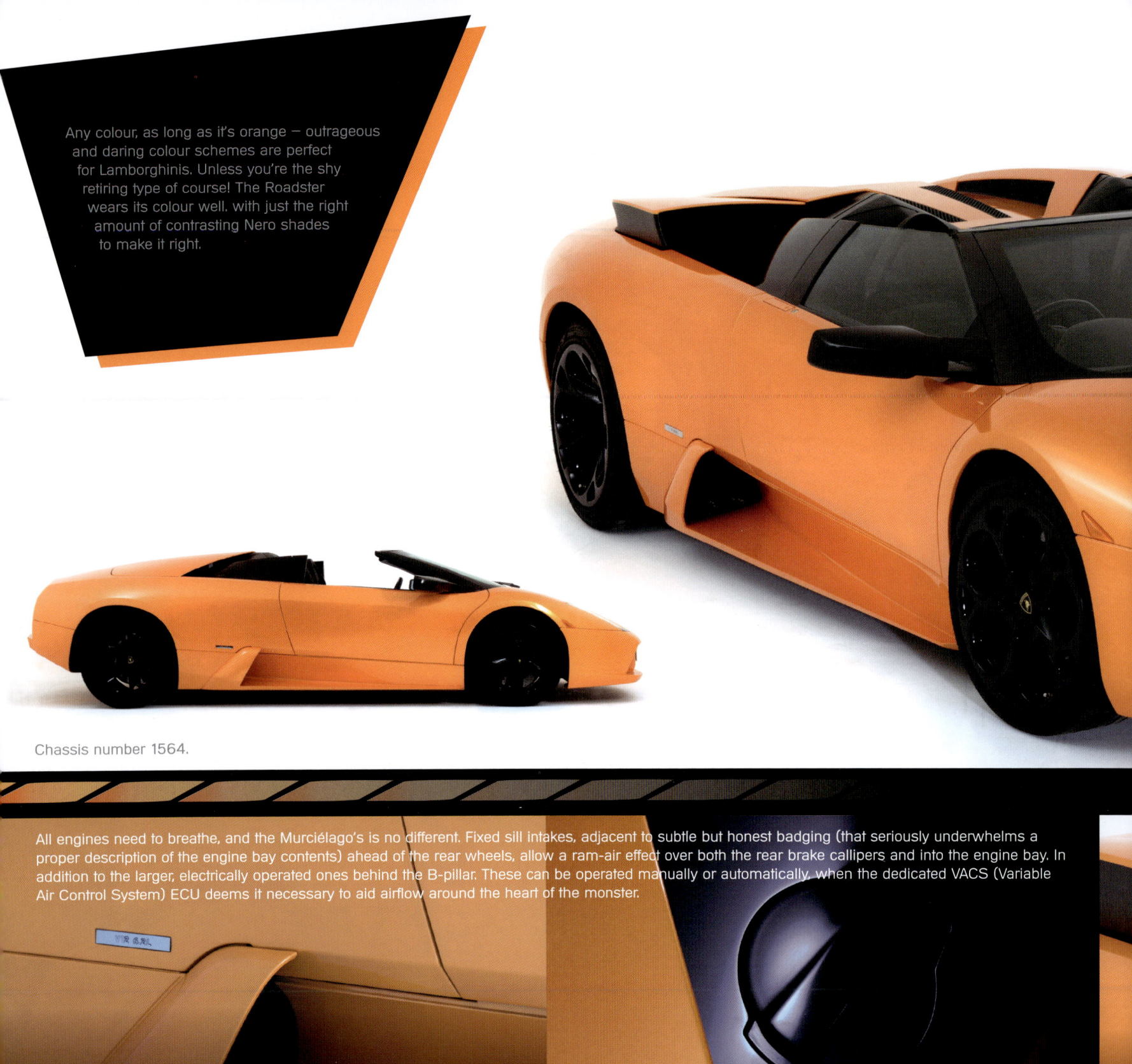

Any colour, as long as it's orange – outrageous and daring colour schemes are perfect for Lamborghinis. Unless you're the shy retiring type of course! The Roadster wears its colour well, with just the right amount of contrasting Nero shades to make it right.

Chassis number 1564.

All engines need to breathe, and the Murciélago's is no different. Fixed sill intakes, adjacent to subtle but honest badging (that seriously underwhelms a proper description of the engine bay contents) ahead of the rear wheels, allow a ram-air effect over both the rear brake callipers and into the engine bay. In addition to the larger, electrically operated ones behind the B-pillar. These can be operated manually or automatically, when the dedicated VACS (Variable Air Control System) ECU deems it necessary to aid airflow around the heart of the monster.

The defining trademark of any longitudinal, mid-engined V12 Lamborghini is the presence of 'guillotine' doors, which on the Murciélago open vertically, adding no width to the occupying volume of the car when opened. A triumph of design, they are often described incorrectly as 'Gullwing' doors in the spirit of Mercedes-Benz, which of course, open upwards and outwards, making life in car parks occasionally tricky ...

(1) With the V12 extracted from the engine bay, one begins to appreciate the task in hand for Lamborghini engineers when taking on servicing work — with the power plant removed, (the pictures here were taken whilst replacing the rear main seal) major components can be accessed more easily. (2) Note the main silencer in the foreground. (3, 4) Ancillaries can be identified here — the most common view seen, tends to be of the cam covers, proudly displaying the Lamborghini name. The manifold is a piece of art in itself, and one half with heat shields in place, can be seen here, its six ports are clearly visible. (5) Once the engine is out of the equation, the tubular cradle of the car and engine mounts are exposed, along with the side heat shields. Without these, the excess heat could damage tyres, the fuel tank and other components. It's also worth noting that, it has to all go back perfectly.

downdraft Weber racing carburettors. With all this, his 60-degree V12 achieved 358bhp at 9800rpm, and 240lb-ft torque at 6000rpm, on 15th May 1963.

Lamborghini, however, was very unhappy with this prototype engine; he wanted a civilised road engine, not a spiky racing engine, and refused to pay Bizzarrini the full second instalment of his fee until forced to do so through court action.

It therefore fell on the shoulders of 26-year-old Giampaolo Dallara, who had been appointed Chief Engineer of Lamborghini at the tender age of 24, to tame this engine. In this, he was helped by 27-year-old Paolo Stanzani. Both these men were later instrumental in the engineering of the Lamborghini Miura, the great grandfather of the Murciélago. It was Stanzani who developed, and saw through to production, the ingenious south-north orientation of the Countach engine, with the drive coming forwards to the gearbox before

descending down courtesy of a drop gear, to then run backwards through a sealed tube in the engine sump, to the rear differential and the final drive. This basic concept lives on, with modifications, in the four-wheel drive Murciélago.

Dallara and Stanzani started the taming process by getting rid of the costly and complex dry sump lubrication system (it only reappeared 38 years later in the 2001 Murciélago VT Coupé. Alterations were also made to the distributors, tappets, valve stems and valve timing, while less costly side-draft Weber carburettors were fitted. Finally, after much work, this modified 513lb engine was able to smoothly deliver 270bhp at 6500rpm and 240lb-ft at 4000rpm, and Ferruccio was comfortable with this engine for his 'perfect grand-touring road car;' the front-engined 350 GTV of 1963.

Iterations of this engine were used in the many front-engined Lamborghinis that followed, and also in a transverse configuration

4

5

in the rear mid-engined Miura, before being used in a daring south-north longitudinal configuration in the Countach, Diablo and finally the Murciélago.

The engine capacity was progressively increased over time, which was only possible because Bizzarrini had designed his engine with an ample cylinder-centre spacing of 95mm. In the LP400 Countach, with a capacity of 3929cc, the engine supposedly produced 375bhp at 8000rpm and 266lb-ft at 5500rpm, progressing through to the LP400 S (same capacity) with 353bhp at 7500rpm and 267lb-ft at 5500rpm. Then there was the LP500 S with 4754cc, and a more honest 375bhp at 7000rpm and 302lb-ft at 4500rpm. A major change took place with the Countach QV of 1985, as the celebrated ex-Maserati engineer, Giulio Alfieri (famed for the Maserati Tipo 61 – the 1961 Birdcage – the 1957 Maserati 250 F, and the 1957 Maserati 3500 GT) introduced four valves per cylinder, and simultaneously lengthened the stroke to give a total swept capacity of 5167cc and 455bhp at 7000rpm and 369lb-ft at 5200rpm.

For the Diablo, Alfieri was joined by former Ferrari racing engineer Luigi Marmiroli, and, at its 1990 debut on the Casino Square in Monte Carlo, the engine in this new model (which had been stretched to 5707cc) produced 492bhp at 6800rpm, and 428lb-ft at 5200rpm. Without going into every variant of the Diablo in this narrative, the general trend in the development of the Bizzarrini engine can be followed through the following examples. The south-north engine orientation afforded relatively easy installation of the new four-wheel drive system in the 1993 Diablo VT. Over time, changes were made to the connecting rods, the big end journals and valve clearance shims, and silicon carbide plating of the bore surfaces was introduced. The 1993 Diablo SE produced 520bhp at 7100rpm, and in the 1999 Diablo GT the stroke was stretched by 4mm to give 5992cc, 575bhp at 7300rpm and 464lb-ft at 5500rpm. The last of the line Diablo, the

Heart of the bull: Extracted from its usual position in the engine bay, one can appreciate just how compact a design the V12 really is. Built for explosive, raw power, controlled and tamed (almost!) by technological advances and dynamic thinking from Lamborghini's exceptional engineering team. This really is a world-class unit.

Heads up — with covers removed, the four cams, engineered to perfection, can be seen, and one can understand how they allow the V12 perfect breath control. Inbetween them, deep below the plug recesses, sparks and fuel mix in a myriad of contained explosions creating the power for a dozen pistons to do their worst!

E-Gear actuators can be seen overlying the gearbox. Note the compact casing.

Cam covers with their associated 'spaghetti!'

6.0 VT maintained the same cubic capacity, but produced 550bhp at 7100rpm and 456lb-ft at 5500rpm.

THE MURCIÉLAGO ENGINE

The Murciélago VT engine is an all-aluminium alloy, 60-degree V12, which carries the internal Lamborghini designation of L535.

This unit was yet another iteration of the Bizzarrini engine of 1963, but on this occasion was developed under Massimo Ceccarani, the Technical Director of Lamborghini at that time.

The final variants of the Murciélago would be the last Lamborghinis to feature this classic power plant, whose own history is so inextricably intertwined with, and has so famously contributed to, the mystique of the Sant'Agata supercar manufacturer.

The original plan for the new Murciélago engine had been to retain the 5992cc capacity of the engine in the final version of the Diablo, but two facts led to the adoption of a larger capacity engine.

Firstly, and despite its more extensive use of composites, the Murciélago weighed 1650kg – 25kg heavier than the Diablo 6.0 – and so needed a commensurate increase in power output, to at least match, but preferably exceed, the power to weight ratio of its predecessor.

Secondly, Ceccarani felt that having a different – and again preferably larger – engine than that in the Diablo would be more attractive from a marketing viewpoint.

Three prototype 6.2-litre engines were built and bench-tested on the factory dynamometers before final approval, and were found to be extremely reliable. In the end, the Murciélago debuted with a 6.2-litre engine, with a power-to-weight ratio of 351bhp/ton, while the Diablo 6.0 exited stage left with 343bhp/ton.

It was a key priority of the L535 to comply with the ever more strict emission standards being introduced around the world, particularly those of its key markets: the United States, Europe and Japan. To this end, several new technological developments were introduced, not only to the powerplant, but also to many other aspects of the Murciélago VT.

The L535 had a bore of 87mm and a stroke of 86.6mm, for a total displacement of 6192cc. With its four camshafts, four valves per cylinder, and a compression ratio of 10.7 (+/- 0.2:1), this engine produced a maximum power output of 571bhp (426kW/580hp) at 7500rpm, and a maximum torque of 479lb-ft (650Nm) at 5400rpm. The rev limiter was set to kick in at 7800rpm.

In its marketing literature, Lamborghini claimed that at just 2000rpm, the Murciélago VT produced more torque than the peak torque achieved by the majority of its contemporary Gran Turismo competitors. This relatively flat torque curve, together with low emissions and better drivability, was made possible by having an electronically-controlled variable geometry air intake system (VIS), and variable valve timing (VVT) during both the inlet and outlet phases, which was again electronically-controlled. The Murciélago also featured 'drive-by-wire' electronic throttle control, to the dismay of some, but this was said to help with emissions, driveability and improved idle speed control.

A major change with the L535 over the engine in the Diablo 6.0 was the introduction of dry sump lubrication – a throwback to the original Bizzarrini prototype engine of 1963. This new engine had an oil scavenger pump that had twice the pressure pump's capacity, and also a deaerator, to remove air and foam from the lubricant oil.

With dry sump lubrication, the engine could now be positioned a full 50mm lower within the chassis, lowering the car's centre of gravity, and thereby improving stability, handling and roadholding. The lower positioned engine also allowed alterations to the car's rear bodywork, which further improved its aerodynamic profile.

The Murciélago VT engine has static ignition with single coils, one for each sparkplug, and the firing sequence of the cylinders is 1-7-4-10-2-8-6-12-3-9-5-11.

Lamborghini, under the stewardship of the Volkswagen Group and its subsidiary Audi, has lost its reputation from the days-of-yore, of making outrageous, unsubstantiated – indeed, almost impossible to substantiate – performance claims. (When first shown at the 1971 Geneva Motor Show, the rumour was put out that the one-off prototype LP500 Countach could exceed 200mph.)

Lamborghini claims that the L535 propels the Murciélago Coupé to a top speed "exceeding 330km/h," and that the 0-100km/h sprint takes 3.8 seconds.

The associated fuel consumption figures for the Murciélago VT released by Lamborghini are:

City	32.6L/100km
Outside city	15.11L/100km
Combined	21.5L/100km

THE GEARBOX

The Murciélago made its 2001 debut with a brand new gearbox; this was the first time in Lamborghini's history that a production car had six forward gears.

The same CEEMA-made gearbox, in modified form, also features in the Pagani Zonda and the Gumpert Apollo. The six forward gears are set in a double-H pattern, with engagement of reverse requiring the gearshift lever to be pressed first downwards, and then to the left and forwards.

The gearshift lever is a tall, thin silver rod, topped with a gorgeous, slightly oversized silver sphere. Set as it is, atop a silver disc embossed with the seven gear positions, which itself sits upon the central transmission tunnel plateau, these two objects are dominant articles of cockpit furniture, and draw the eye towards themselves, particularly in a dark-hued cabin.

The primary and secondary shafts in the gearbox of the Murciélago are each mounted on three bearings, rather than the two in the Diablo, to help with rigidity and stability of the transmission system. New double and triple cone synchronizers have been introduced, as well as changes to the control linkages, to improve the gear change feel and reliability.

Gearbox lubrication in the Murciélago is through a pump, which is located within the gearbox itself.

This long-serving CEEMA box can be somewhat recalcitrant, but is a significant improvement over that in the Diablo. As with all cars of this ilk, patience and mechanical sympathy are essential, and with the gearbox oil warmed up – which can take a considerable time – the shifts become more slick. Even so, a positive movement on the part of the driver is required to change gear, but this is surely part of the joy of a Murciélago – the sensation as one gear is disengaged, the stick fed past neutral, and then another one engaged with real mechanical feel. Clutch depression too requires some muscle, but less so than in the Diablo, and much less so than in the notorious Countach – but, again, this is something positive, something to be desired. A Murciélago with a buttersmooth gearlever movement, and a city car clutch weight would be the lesser for these 'improvements,' and, in the case of the Countach, the loss of the gear change and clutch weights would be tantamount to robbery. In the Murciélago, the clutch slave cylinder has been fitted in line with the clutch release bearing axis to make clutch depression easier.

The standard Murciélago 6.2 VT gear ratios are:

First gear	1:2.941
Second gear	1:2.056
Third gear	1:1.520
Fourth gear	1:1.179
Fifth gear	1:1.030
Sixth gear	1:0.914
Reverse gear	1:2.529

The Murciélago 6.2 VT could also be ordered with an optional close-ratio gearbox, with the following ratios:

First gear	1:2.941
Second gear	1:2.056
Third gear	1:1.520
Fourth gear	1:1.259
Fifth gear	1:1.061
Sixth gear	1:0.943
Reverse gear	1:2.529

E-GEAR

The optional E-Gear transmission system available on the Murciélago is a robotised manual transmission system, which is otherwise also known as an automated manual transmission system, or AMT. Initially used in Formula One racing cars, Magneti Marelli first introduced this system for road car use in the Ferrari 355 F1 in July 1997.

The E-Gear option allows one of a number of different driving modes to be chosen by the driver. Firstly, the driver could choose the fully automatic mode, where a computer decides when a higher

Pressing buttons: E-Gear gives the driver optimum performance from the gearbox – beats cramps in the calf muscles too, with only light touches on the paddle shift levers required.

Lamborghini has become something of the expert on composite technology within the Volkswagen Group. It established an Advanced Composite Research Centre at the factory in 2007, with an engineering division and a manufacturing division. Lamborghini claims to be the only car manufacturer in the world to have an in-house facility that can take a composite product idea through the concept, manufacturing and finally, if necessary, repair stages, without recourse to outside help. Lamborghini also set up a collaboration with Boeing in 2007, focussing on material crashworthiness, modular construction processes, and the repair of damaged composite structures.. This cooperation in material sciences between Lamborghini, Boeing and various universities and research institutes from around the world was further cemented by the establishment of the Advanced Composite Structures Laboratory in Seattle, in 2014. Lamborghini's Sesto Elemento prototype is an example of this technology in action; not only the monocoque, but also the shock absorbers, wheels and transmission shaft in the Sesto are made of carbon fibre. Another example is a material also developed by Lamborghini's composite division, called Carbonskin – a flexible carbon fibre matrix suitable for use in a car's interior that is pleasing to the eye and touch, durable, and weighs 65 per cent less than leather, and 28 per cent less than Alcantara.

Sesto Elemento, (or 'Sixth Element') Shown at the 2010 Paris Motor Show. The concept is named after the atomic number of carbon — extensively used in this design showpiece that confirms Lamborghini's ongoing commitment to pioneering design. For example, only Lamborghini would dare to 'reverse' door mirrors, and make them look not only correct, but stunning, too.

Aftermarket carbon-fibre roof, Geneva 2007. Which would you choose, the hard or soft option?

or lower gear should be selected, and then actually carries out this decision. Alternatively, the driver could choose to manually select the gear using the steering wheel column-mounted paddles; within this second mode, the driver can further select between three different driving programmes: Normal, Sport and Slippery-Road.

In the default Normal programme, the car will automatically change up to the next higher gear when the engine speed reaches 7450rpm.

In the Sport programme, there is no automatic up-shifting, and the engine speed is maintained at the rev-limit once it has been reached. The time required for the system to execute a gearshift is also quicker in Sport mode, and particularly so if the accelerator pedal is floored and the engine speed is above 6500rpm.

In the Slippery-Road programme, the system uses softer pickup maps for first, second and reverse gears, and automatically engages the next higher gear as soon as the engine speed reaches 3200rpm. This programme also sets a lower threshold for the Traction Control System to cut-in, and overrides the Sports programme should it have been selected previously.

While the gearbox components of a manual and E-Gear Murciélago are almost identical, in the E-Gear variant, both the clutch operation and the gear selection operation are electrohydraulically controlled. There is therefore no need for a clutch pedal. An electronic control unit dictates the clutch disengagement, gear engagement and clutch re-engagement movements, through specific high-pressure hydraulic actuators. The classic gearstick and open gate is replaced in the E-Gear variant by two alloy paddles behind the steering wheel, and by a closed alloy disc on the transmission tunnel carrying two control buttons (Sports and Slippery-Road) for the E-Gear system.

In the Murciélago, the paddles are fixed to the steering column, and a pull on the right paddle engages the next higher gear, while a pull on the left paddle engages the next lower gear. Neutral is engaged by pulling on both paddles simultaneously, while Reverse is engaged by pressing a button on the lower Dashboard column.

The E-Gear option also allows for what Lamborghini calls a "Burn-Off" start. This is an electromechanical procedure to get the fastest possible acceleration from standstill that the car is capable of. To enable a Burn-Off start, the vehicle needs to be stationary, with first gear engaged, the traction control system deactivated, and the Sport programme selected. If all these conditions are met, then on flooring the accelerator pedal, the car enters into a special operating mode in which the engine speed is automatically increased to the point where the maximum possible torque is available, and the clutch is then automatically released to best utilise all this available torque.

It requires some finesse to drive an E-Gear Murciélago smoothly, especially at crawling speed, when the clutch can be heard spasmodically catching and releasing. Once at speed, the E-Gear system is much better, especially when the more aggressive Sports mode is engaged. The low speed jerkiness is replaced by smooth, quick gearchanges. Although nowhere as refined or as fast as the latest dual clutch systems, the now antiquated single clutch robotised

Key players

At the 2001 Sant'Agata Murciélago launch, I had the pleasure of meeting some notable ex-Lamborghini personnel:

With Horacio Pagani, ex-Lamborghini engineer, founder of Modena Design, Countach Evoluzione visionary and Zonda designer.

With Paolo Stanzani – Chief technical officer for Lamborghini, also assisted Romano Artioli with the relaunch of Bugatti.

With Loris Bicocchi, chassis guru/test engineer with Lamborghini, Bugatti, Zonda, KTM and Konigsegg, to name but a few.

manual system in the Murciélago still has its own honest charm, with a more genuine mechanical feel than the newer systems, while the fixed paddles are beautifully positioned, and a delight to use.

LUC DONCKERWOLKE

The word Murciélago means 'bat' in Spanish, and the body of this new model had, for the first time in Lamborghini's production car history, movable air intakes on the upper flanks of the rear bodywork. With these intakes in the raised position, the Murciélago has some resemblance to a bat with its wings outstretched. The Murciélago's bodywork represented a major change for Lamborghini in two other respects: this was the first clean sheet car designed by Luc Donckerwolke for Lamborghini, the three previous mid-engined Lamborghini flagships having been designed by the legendary Marcello Gandini; and secondly, there was a very substantial increase in the use of carbon composites in the Murciélago compared to its predecessor.

When the Volkswagen Group acquired Lamborghini in September 1998, it was worked out that Lamborghini would need to sell 1500 cars a year to remain viable. Lamborghini had established rivals like Ferrari and Porsche, but there were now also newcomers like Pagani, which introduced the Zonda in 1999, and Koenigsegg, which was established in 1994 and delivered its first production car in 2002. To keep up, Lamborghini needed a new flagship to replace its aging Diablo, which had made its debut in 1990, and it needed it fast.

As Lamborghini passed through the hands of Chrysler and MegaTech, and finally into those of the Volkswagen Group, a number of styling houses put in proposals for the Diablo replacement. These included proposals from Bertone and Giugiaro's Italdesign, as well as the Lamborghini Canto prototype by Carrozzeria Zagota. The Canto featured massive fixed, ellipsoidal rear air intakes that sat on top of the rear wings, and which – although dramatic – looked like tacked-on afterthoughts.

Volkswagen Group Chairman Ferdinand Piech hated the Canto, and rejected the proposal on the basis that it did not express the DNA of Lamborghini, and that it was not a worthy successor to the Miura, Countach and Diablo.

At this stage, the Volkswagen Group asked IDEA, Bertone and Heuliez to rework the Canto design, but simultaneously decided to establish an in-house Lamborghini design studio at the factory complex in Sant'Agata. This in-house studio was headed by Luc Donckerwolke, whose only previous road supercar design experience was in restyling the Diablo into its VT 6.0 variant.

The son of a Belgian diplomat, Luc Donckerwolke was born in Lima, Peru in 1965, has lived in at least 13 different countries, and speaks eight languages. He studied industrial design in Brussels, and

transportation design in Vevey, Switzerland, before starting his design career in 1990 with Peugeot in France. After two years with Peugeot, he moved to Audi Design in Germany for another two years, before moving to the Škoda Design Centre, where he again remained for two years. After this he returned to Audi, where he worked on the Audi R8 Le Mans Racer, and the Audi A2 concept and production cars. Through his work on the A2, he gained a reputation for innovation, and for being ahead of his time.

Donckerwolke tells an interesting story of how he gained his promotion within the Volkswagen Group empire, to become the head of design at Lamborghini in 1998. Peter Schreyer was at that time the head of design at Audi, and Luc's immediate boss. Donckerwolke says that he was summoned into Schreyer's office, and told that he was going to be offered what was described to him as an exceptional job opportunity. He was only told that the job was not based in Germany, that it would be a difficult challenge, and that the details of this job could not be revealed to him until and unless he accepted the opportunity blind, and that – either way – he needed to give an immediate answer as to whether or not he would be prepared to take on this new challenge. When he accepted the deal, and was told that the job was at Lamborghini, Luc described it as an unexpected dream come true.

Having successfully overseen the restyling of the Diablo VT 6.0, Donckerwolke's proposal for the Diablo replacement was accepted by the Volkswagen-Audi-Lamborghini board. Donckerwolke had already studied and rejected the idea of re-working the Canto design, and decided that a clean sheet approach was the only way to design a worthy successor for this famed lineage.

Donckerwolke said that he wanted to take a purist approach to the new model, and that it had to have within it the DNA of the Miura, Countach and Diablo. To this end, he started off by carefully studying both the silhouette, as well as the styling details within each of these three cars.

Donckerwolke also tells a lovely anecdote about being introduced to the legendary Lamborghini test driver, Valentino Balboni, who said to him, "You are the man who designed the engine cover." Donckerwolke answered to this in the negative, further explaining that he had designed the whole car. Balboni apparently quickly replied saying, "That's right. We did the engine, and you did the bit that covers it."

CARBON FIBRE CLOTHING

Although Lamborghini's experience with the automotive use of composite materials reaches back a long time, the Murciélago range used a lot more composites, and in particular carbon fibre composites, than any of its predecessors.

Lamborghini itself suggests that it was the famous hands-on engineer-turned-general manager, Giulio Alfieri, who recognised the need for a lighter and stronger material than aluminium. In 1982, he obtained funding from the European Economic Community for the design and construction of the experimental Lamborghini Countach Evoluzione of 1987, which had a 100 per cent carbon fibre composite body, and was variously said to be between 390kg and 500kg lighter than a standard production Countach. The Evoluzione was never intended as a production car, nor even as a prototype car; it was purely a design test bed. The legacy of the Evoluzione was the ever-increasing use of composites, as the Countach QV evolved into the Countach Anniversary, before transmuting into the Diablo, and then into the Murciélago. Horacio Pagani, of Zonda and Huayra fame, was instrumental in the development of the Evoluzione, and no doubt took and utilised knowledge from this experimental project for use in his own carbon-rich supercar.

The bodywork of the Murciélago is made of carbon fibre (excluding the roof and the external panels of the doors, which are primarily steel). Parts of the frame also feature carbon fibre structural parts, and, in the Roadster variant, the lattice cage over the engine could also be specified in carbon fibre as an option.

Carbon fibre was first produced by Joseph Swan, a British chemist and physicist, in 1860, for use in incandescent light bulbs. Sadly, this was commercially impractical at that time, and the venture was a failure. Thomas Edison, a prolific inventor, produced the first practical and commercially successful all-carbon fibre incandescent light bulb filament in 1879. Over the next 70 years various companies, particularly in the United States and Japan, attempted to discover a way of efficiently producing industrial quantities of carbon fibre, while also achieving the high strength and stiffness that carbon fibre is theoretically capable of.

Carbon fibre is made up of long, thin strands, which themselves are almost exclusively composed of carbon atoms. The carbon atoms are bonded together into microscopic crystals, which in turn are aligned along the long axis of the fibre to form filaments. Each carbon filament is a continuous cylinder with a diameter of about five to ten micrometres. Thousands of these filaments are twisted together to form a yarn, which can then be used by itself, or woven into a fabric. Once the carbon fibre yarn or fabric is impregnated with a plastic polymer resin and baked, it forms a carbon-fibre-reinforced-polymer (commonly abbreviated to carbon fibre), which has an exceptionally high strength-to-weight ratio, and is also very rigid. There are a number of different autoclaving techniques available for curing carbon fibre, but all essentially attempt to achieve a given desired fibre to resin ratio, with the elimination of resin voids, which weaken the carbon fibre structure.

The carbon fibre fabric is carefully laid up in an almost surgically sterile environment, impregnated with resin, placed under vacuum, and then pressurised in an autoclave while going through a heat cure cycle. The high pressure and temperature within the autoclave (typically a nitrogen atmosphere, set at about seven bar, and running at between 120 and 230 degrees Centigrade, with curing times that can range from 90 minutes to 12 hours) can achieve a resin void content of less than two per cent, which is acceptable for aerospace structures and Lamborghini.

Carbon fibres are classified according to the tensile modulus of the fibre, which is measured in pounds of force per square inch of cross-sectional area. The strongest carbon fibres are ten times stronger than steel, while simultaneously being five times lighter. Carbon fibre also has excellent fatigue- and corrosion-resistant properties. The main drawback to carbon fibre has, traditionally, been its high production cost. More recently it has become widely known that carbon fibre is wasteful to produce, and difficult to recycle.

Nevertheless, industry loves the exceptionally high strength-to-weight ratio and rigidity of carbon fibre, and the use of carbon fibre globally has been increasing at a rate of about ten per cent year-on-year. Carbon fibre is particularly used in aerospace applications, military engineering, wind turbine blades and, of course, high-performance sports and racing cars. Once the preserve of ultra-exclusive and hyper-expensive cars like the Pagani Zonda, carbon fibre tubs are now available in slightly more accessible cars like the KTM X-Bow.

The Austrian-built, Audi-gearboxed and -engined KTM X-Bow has a convoluted relationship with Lamborghini, which transcends the Audi connection. Giampaolo Dallara – who joined Automobili Lamborghini as Chief Engineer at its inception in 1963, and who was so instrumental in the development of the Miura and the Countach – later set up what is now the legendary Italian racing car manufacturer, Dallara. KTM claim the X-Bow as the world's first large-scale production car with a full carbon composite monocoque, and Dallara was intimately involved in the development of that monocoque, as well as the car's aerodynamics and its suspension set-up. Another person who was closely involved with both marques is the hugely respected chassis guru, Loris Bicocchi. Bicocchi was actually born in Sant'Agata, and was schooled as a test driver at his local youth training centre: that little-known neighbourhood establishment, Automobili Lamborghini SpA, under the tutelage of Valentino Balboni. He had started off working in the Research and Development warehouse at Lamborghini in 1974, but within a year had wangled his way into being employed as a mechanic. Paolo Stanzani, another legendary Lamborghini figure, enticed Bicocchi to move with him to work on the Bugatti EB110 project, following which Bicocchi was responsible for the chassis set-up of the Pagani Zonda, Koenigsegg, Bugatti Veyron, and – most recently – the Bugatti Chiron. KTM recognised his engineering talent, as well as the marketing value of having someone of Loris Bicocchi's stature responsible for the chassis set-up of the X-Bow. The result was a success, in that this race-biased road car was deemed very civilised and well suited for street use, if a little less raw than some would have liked. Bicocchi also holds a special appeal for me, as he told me in 2001, at the Murciélago launch event, that he almost certainly did the pre-delivery inspection and test drive on my Countach 5000 QV, chassis 12399.

THE MURCIÉLAGO INTERIOR

The guillotine doors of the V12 mid-engined Lamborghini flagships are arguably the single most dramatic aspect of these cars, and have rightly become their signature feature. Even the dramatic wedge profile is overshadowed by these upswinging doors.

With the Countach, the door release mechanism button is hidden within the NACA duct, and requires the uninformed to bend low to find it. With the Murciélago, the release mechanism advertises its presence through a long rectangular lever with a round indent at one end, which sits flush with the top of the door panel, just beneath the side window glass.

Pressing down on the indent with one's index finger causes the other end of the lever to stand proud of the bodywork, and one's other fingers are now naturally positioned to grip the other end of the lever. Pulling the lever upwards releases the locking mechanism, and allows the front hinged door to be moved skywards. Although assisted by gas struts, it is best to help the doors in their ascent by pushing gently on their lower edges once the doors have opened to about 30 degrees, so as not to exert undue force on the door lever.

Compared to the Diablo, the door opening angle has been increased by five degrees, and the sills – which now carry an embossed Murciélago logo – are 25mm lower, to allow easier and more graceful ingress and egress.

The Countach models (except the Anniversary) have a lovely rounded leather-bound sill cover, which allows the driver and passenger to sit on it, and then literally allow themselves to fall sideways and backwards, and thereby glide smoothly and relatively elegantly into their respective seats, before curling their feet and legs into the foot wells. In the Murciélago, however, the sill cover is more right-angled, which disallows the aforementioned method of entry, which is a bit of a disappointment.

What is *not* a disappointment is the sight of the beautiful Murciélago cockpit that is revealed through the open doors. There is leather everywhere, and the coloured piping seen so commonly in the Countach has been replaced by coloured stitching in many Murciélagos.

Even after almost 30 years, opening the Countach's door releases a very special waft of leather – just glorious, and, in my limited experience, unique – but sadly the leather of the Murciélago does not give off this special smell. This loss is tempered by the excellent fit and finish of the interior appointments; there are no readily identifiable bits lifted from the Audi parts bin, and most of the major furnishings appear to be bespoke.

The interior was designed in-house by Audi designer Ralph Kluge, who had previously worked on the Audi Rosemeyer concept car, and it is in many ways quite sombre and conventional, in contrast to Luc Donckerwolke's dramatic exterior.

The arrow-shaped seat headrests appear to be free-floating, but the seats are otherwise orthodox. The seatbelts are unusual, in that the male buckle connector is attached to the central tunnel, and the belt is pulled downwards and laterally, towards the outside of the car, to meet with the female buckle socket, which is also where, on the driver's side, the handbrake lever resides. The central transmission tunnel descends from between the seats, carrying within it both a shallow oddments compartment and an ashtray, before reaching a plateau.

This flat area has, in the back row, the fuel cap release button, the hazard flashers and the external mirrors retractor button. In the middle row are the electric side window controls, and the metal ball-topped gearshift lever in manual transmission cars, or the E-Gear system controls in the robotised transmission cars. Finally, in the front row is the front lift button, the traction control disabling button, the VACS control button, and the controls for adjusting the position of the door mirrors.

This flat area then meets a vertical panel housing the audio and ventilation controls (and the optional navigation system), which itself then merges with the dashboard.

On the passenger side there is a grab-handle, as in the Miura, so that the passenger is not overly flung about the cabin when the Murciélago's power and torque conspire with its four-wheel drive system to produce the highest g-forces possible.

There are four round chrome-ringed air vents, and a still modern-looking angular dash binnacle, although the speedometer and the rev-counter are both traditional.

The small Momo steering wheel eschews the modern trend of flat-bottomed wheels, in favour of a perfectly round, rather colourless wheel. In my opinion, it has the perfect steering wheel diameter and rim thickness (which, when driving, contributes to the delightful weight of the steering, in this power assisted car). The addition of a small,

Simple and unadorned, but exquisitely appointed, from the perforated leather on the driver's side, to the restrained and minimalistic styling of the switchgear, camouflaging the amount of technology that lies beneath. The Murciélago's interior space is designed with the focus on the driver, of course!

easily removed, high-quality, coloured, slightly raised, 3D Lamborghini logo sticker, really lights up both the steering-wheel and the cabin.

Immediately behind the steering wheel are the E-Gear paddles, relatively small by modern standards, but with a satisfyingly positive action. A pull on the right paddle brings the next higher gear into play, and a pull on the left paddle does the converse.

The wiper and light/turn control stalks are on the right and left sides respectively, and the steering wheel adjustment locking/unlocking lever is on the undersurface of the steering column (the steering wheel adjusts for both reach and rake). On the side of this column is the ignition key slot.

Towards the door, level with the ignition key slot on the driver's side is a row of five discreet buttons, the outermost and most important of which, is the one that selects the reverse gear in the E-Gear format. The other four buttons are the controls for the cockpit roof light, the parking lights, and the front and rear fog lights.

The angled dash binnacle houses two primary dials: the speedometer and the rev-counter, both with clear white on black markings. There are four secondary dials: the engine coolant temperature gauge, the fuel gauge, the engine oil temperature gauge, and the engine oil pressure gauge. There are also either two or three additional dash displays, depending on the type of transmission in that particular car: the drive information display, the odometer/trip meter display, and – only with the optional E-Gear system – a display for the currently selected gear. There are also 34 (!) dash warning and indicator lights, which indicate the activation of various functions, or the development of various faults.

Maybe Lamborghini should also introduce two further dials in the future; one indicating the amount of forthcoming bother in getting the car to the dealership for the repairs, and another to indicate just how wallet-wilting the repairs are going to be.

In the footwell, there is a footrest for the left foot, and both it and the accelerator pedal both have a stylish metal surface finish.

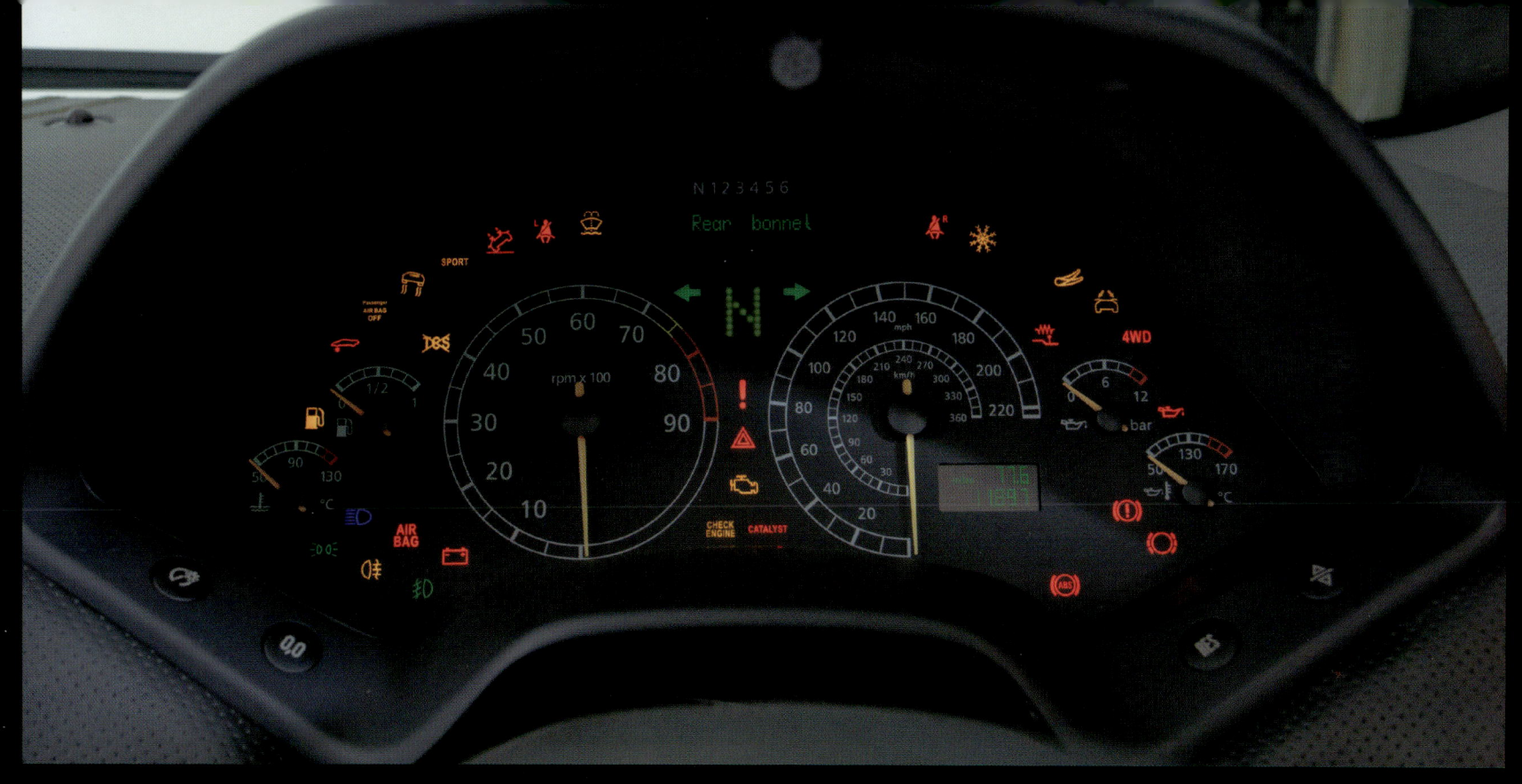

The cockpit is not snug, with both the windscreen base and the passenger door appearing to be a Pacific Ocean away from the seated driver. While not claustrophobic – and while larger drivers can be more comfortably accommodated in the Murciélago than in any of its mid-engined predecessors – there is relatively little driver/passenger space or storage space within the cockpit, considering the length and width of the car. This is a trait that the Murciélago shares with many of its mid-engined supercar brethren.

Audi's influence shows through in the cabin ergonomics, which are much superior to that in the Diablo, let alone the Countach.

The view back through the rear view and door mirrors are poor, with large blind spots; pretty much the accepted norm for mid-engined supercars of that era. The door mirrors largely serve up a view of the voluptuous rear wings, while the Miura-like slats of the engine lid further compromise the already slit-like view directly backwards.

The heating, ventilation and air-conditioning all work pretty much perfectly, and all the secondary controls have a proper and consistent weighting, which makes them delightful to use.

The inner surfaces of the doors are also leather-lined, and feature two audio speakers, a window raising and lowering button, the inside door release handle, and some storage space.

THE FOUR-WHEEL DRIVE VISCO-TRACTION SYSTEM AND DIFFERENTIALS

The permanent four-wheel drive system in the Murciélago uses a derivation of the system found in the Diablo, and features a central viscous coupling, which Lamborghini terms the Viscous Traction System.

Self-blocking differentials allow 25 per cent limited slip at the front, and 45 per cent limited slip at the rear, which allows any excess torque from the primary axle – the rear axle – to be transferred to the secondary axle at the front. In so doing, this system approximates an active traction control system, and attempts to maintain the optimal traction possible whenever the car is in motion. The front differential has a ratio of 1:2.529, and the rear 1:2.933.

There is, in addition, a completely separate, electronically-managed traction control system that works by limiting engine torque in extreme driving situations. This reduction in torque is achieved by electronic intervention on the throttle through the drive-by-wire system, as well as by electronic intervention on the fuel-injection and ignition systems.

Whenever the traction control system is automatically activated, the driver is alerted to its activation by a flashing yellow warning light on the dashboard. However, I suspect that by this stage most drivers might be too busy managing the steering wheel and the accelerator pedal to spend much time admiring this pulsating yellow icon.

There is also a traction control cut-out switch located on the plateau of the central transmission tunnel. It would seem that this facility would only be appropriate on track.

To purposefully digress, to me, the Murciélago is spectacularly unsuited for the track. My two track-day cars (Brooke 260 RR and KTM X-Bow R) are the antithesis of the Murciélago, which I feel was designed to be a cross between a Gran Turismo car and a super sports car. The Murciélago, for all its many positive features, lacks the following attributes which I think are hugely desirable in a track car: foremost, a low kerb weight; secondly, having minimal or no electronic aids, so as to maximise driver involvement and pleasure (à la the mantra of Colin Chapman, of Lotus fame: "Simplify, then add lightness." He went on to elaborate, "Adding power makes you faster on the straights, subtracting weight makes you faster everywhere"); thirdly, being narrow; fourthly, having good all-round visibility; fifthly, using relatively cheap and accessible consumables, particularly brake pads, brake discs, clutches that can be replaced without having to remove the engine, etc; and finally, cheap, readily obtainable and easily bolted-on/off body panels in case of accidents.

To return to the technicalities of the Murciélago, switching off the traction control, courtesy of the central console button, lights up the same yellow traction control icon on the dash, but now it stays on permanently, rather than flashing intermittently.

THE CHASSIS

The Murciélago is the last mid-engined V12 Lamborghini to use the classic Superleggera chassis set-up, and, to the romantic in me, this is a substantial loss.

I readily accept that a carbon fibre monocoque is superior to a spaceframe chassis in terms of weight-saving and stiffness, with all the attendant dynamic advantages that these two key attributes bring, but driving an old, naturally-aspirated, internal combustion petrol supercar is about so much more than pure speed, comfort, handling and emissions; there is also a sense of history to these cars, which allows an enthusiast to appreciate them on a different level, and which allows one to appreciate their bespoke nature, even when they are standing still with the engine turned off.

The word Superleggera is Italian for super light, and, while this chassis concept is commonly associated with Carrozzeria Touring, it was actually invented by Charles Terres Weymann. Weymann was a Haitian-born (to a Haitian mother and American father), France-residing, aeroplane-racing pilot, 1928 Le Mans team manager/owner, and aircraft designer, who also patented an automatic clutch system.

Felice Bianchi Anderloni (a lawyer) and Gaetano Ponzoni founded Carrozzeria Touring in Terrazzano di Rho, near Milan, in March 1926, and started constructing cars, having licensed Weymann's idea of fabric covering a lightweight frame.

In 1936 Carrozzeria Touring patented a development of this idea, in which alloy panels replaced the original fabric. The Superleggera system uses a lattice of lightweight, small diameter metal tubes to form the framework of the car's body shape (the spaceframe), which is then covered by thin alloy body panels, which in turn further contributes to the strength of the frame.

The Countach, Diablo and Murciélago all use a spaceframe, the principal difference being that the Marchesi-built Countach spaceframe uses round steel tubing which is much more difficult, more expensive and more time-consuming to weld, than the square tubing found in the later cars.

In the Murciélago, the chassis is made up of high tensile strength steel tubing, with further strengthening achieved through structural carbon fibre elements that are attached to the spaceframe tubing with either steel rivets or adhesives.

A key objective during the development of the Murciélago chassis was to further increase the torsional rigidity of the spaceframe, as this would directly improve the car's roadholding and handling, enhance driver and passenger comfort through better suspension control, and reduce vibration and noise transmission into the cockpit area. To this end, the Coupé has a structural steel roof, and a floorpan made of carbon fibre that is attached to the lower spaceframe.

Through the above modifications, the Murciélago chassis achieves a highly acceptable torsional rigidity in excess of 20,000Nm/degree.

The new Murciélago spaceframe allows the front shock absorbers and front suspension elements to be moved further forwards, freeing up more space in the footwells, which in turn allows for better pedal placement. The wheelbase of the Murciélago has also been increased by 15mm over the Diablo, which further helps stability.

THE SUSPENSION

The 1650kg Murciélago 6.2 VT Coupé has a 48 per cent front and 52 per cent rear weight distribution, which is carried on a classic, independent, front and rear double wishbone suspension set-up.

There is one hydraulic shock absorber with electronic damping control per side on the front, and two such shock absorbers per side on the rear. Each shock absorber is shrouded by a coaxial spring, and this spring-shock absorber combination is attached to the chassis through Flanbloc bushes. There are anti-roll bars front and rear, and the suspension also features 'Antidive' technology at the front, and 'Antisquat' technology at the rear, to keep the car as level as possible during hard braking and acceleration respectively.

Particularly useful for city-dwelling owners is the optional electrohydraulic front lifting system. There is a button on the plateau of the transmission console that, on activation, raises the front of the car by about 40mm, protecting the lower part of the expensive carbon fibre bumper from kissing whatever elevated road surface obstacle which might want to get too intimate with it. This front lift system can only be activated with the engine running, and when the car's speed is less than 70km/h (or for Japanese market cars, less than 30km/h). The car is lowered back to its normal height by pressing the same button again. Owners are advised not to leave their cars with the front suspension elevated for extended periods, as this places undue strain on the shock absorber seals, which can reduce their useful lifespan.

THE ACTIVE REAR SPOILER

The Murciélago features a three-position, electronically-controlled, automatic rear spoiler, which helps the suspension keep the car stable, particularly at high speed, without unnecessarily carrying the penalty of additional drag.

The rear spoiler sits flush with the bodywork until the car reaches a speed of 130km/h, at which point the rear spoiler automatically raises itself to a position of 50 degrees from horizontal.

As the car's speed climbs, it remains at this angle until a speed of 220km/h is reached, at which stage the spoiler rises further, to its final position of 70 degrees from horizontal.

WHEELS AND TYRES

At its 2001 debut, the Murciélago VT Coupé came with new multi-piece 18-inch aluminium alloy wheel rims with concealed valves. These five-hole wheels had more than a passing resemblance to the OZ wheels of the later Countach cars, although in the Countach the rear wheels in particular were deeply dished in a concave pattern, while in the VT they were slightly convex.

These wheels, albeit a little fussy in design, give real presence to the car, not least by virtue of their size, at 8.5x18in at the front, and 13x18in at the rear. These wheels were designed to improve air circulation to the brake discs and pads, and so help reduce brake fade.

Lamborghini recommends Pirelli P Zero Rosso tyres, and the standard fitment tyres measure 245/35 ZR 18 at the front, and 335/30 ZR 18 at the rear. The recommended cold inflation pressures for normal use are 2.6bar for the front tyres, and 2.5bar for those at the rear, rising to 3.2bar for both front and rear tyres for use at speeds above 290km/h.

THE BRAKING SYSTEM

The Murciélago VT had an advanced braking system locked in as a key design priority. It has ventilated, cross-drilled brake discs at each corner, measuring 355mm at the front, and 335mm at the rear, together with four piston brake callipers.

There are two totally independent hydraulic braking systems, one for the front axle brakes and one for the rear axle brakes, and this is supplemented by a four-channel ABS system, and dynamic rear proportioning. To increase safety margins, the braking system is also integrated with the traction control system. The dynamic rear proportioning system has been refined over and above the system in the Diablo, so that the optimum braking force is distributed between the front and rear wheels at all times, and in all conditions.

The TRW four-channel ABS system is able to monitor and control each wheel independently, and consists of four wheel speed sensors linking in to an electrohydraulic control unit, with a dedicated microprocessor. Each of the four hub flange-mounted electromagnetic speed sensors continually sends signals to the microprocessor, which calculates the speed and acceleration of each individual wheel, and so is able to detect any wheel's tendency to slip. If any such slip is detected, the microprocessor can modulate the brake pressure in the

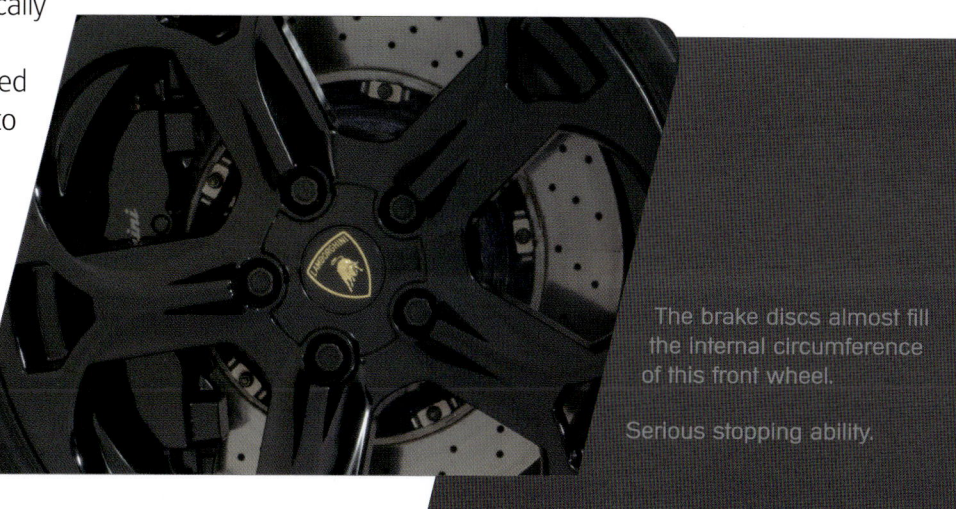

The brake discs almost fill the internal circumference of this front wheel.

Serious stopping ability.

corresponding brake line to stop the slip. If the ABS system breaks down, the basic hydraulic system will continue to work as normal, and a red warning light will come on instantly on the dash display, to warn the driver of this safety critical malfunction.

THE ELECTRONIC MANAGEMENT SYSTEMS

Progressively more stringent government legislation on car safety and emission standards, as well as owner demands regarding performance, reliability and comfort, are best met through the use of sophisticated electronic diagnostic, control and management systems.

In the Murciélago, the Electronic Management system is made up of four principal units, which together control the following functions:

The multipoint, sequential fuel-injection system
The ignition management system
The variable valve timing management of both the inlet and exhaust valves
The variable geometry intake management system
Detonation management
The electronic throttle, drive-by-wire, system
The traction control system
The variable air intake cooling system
The rear spoiler deployment and control system
The OBD II onboard diagnostic system
The Lamborghini LDAS diagnostic system
The three-way catalytic converter and oxygen sensor emission management system

The four principal units, each of which use 32bit/20Mhz microprocessors, are comprised of:

Two Lamborghini LIE management units, each of which controls one of the two engine cylinder banks.
One Lamborghini GFA auxiliary function management unit that controls 'body' functions.
And one Lamborghini Power Motor Control satellite control unit, which also controls 'body' functions.

THE STEERING

The Murciélago's steering is operated through a rack-type system, with hydraulic power assistance. I loved just about every aspect of my long-standing 1994 non-airbagged Ferrari 355 GTB, but really disliked its overly-light assisted steering. As a primary control, the steering is so important to the pleasure of driving, and this was probably the key driving reason that I swapped the 355 for a Murciélago.

Keeping in mind that the Murciélago has permanent four-wheel drive and power-assisted steering, the steering weight and feel are excellent. This is partly due to the lovely three-spoke leather Momo steering wheel, whose diameter and rim width I find ideal. Less ideal is the turning circle, which has a minimum diameter of 12.55m. There are about three steering wheel turns from lock to lock.

SAFETY

From a safety aspect, the steering column is collapsible, and there is a 60-litre airbag mounted within the centre of the steering wheel. This driver's side supplementary restraint system is mirrored on the passenger's side by a 120-litre airbag that is hidden under the dashboard in front of the passenger's seat. Any malfunction of the airbag system is flashed up in red on the dash display. Lamborghini clearly states that the airbag system does not replace the safety belts, which should be used during the course of every journey.

Lamborghini also says that the airbag system is designed to operate in moderate and severe frontal and near-frontal collisions. Airbag inflation will only occur above a set threshold speed of between 7mph to 15mph, but the angle of impact and the deceleration values are also taken into account by the airbag control unit (which is located within the transmission tunnel), which then makes a split-second decision as to whether or not the airbags should be deployed. The three-point cross-abdominal type safety belt system also features pyrotechnic belt tensioners on the buckle-side of the seatbelts.

VISIBILITY

Visibility in the dark is provided by Bi-Xenon headlights on both high and low beams, using Hella 12v (all the bulbs run on 12 volts), 35w bulbs. The front fog light bulbs are 55w bulbs.

The tail light bulbs are rated at 5w, the brake lights, reverse light and the rear fog light 21w, and the twin rear number plate bulbs 5w. The front and rear turn signal bulbs are 21w, and the side turn signal bulbs are 4w.

In the rain, the single arm pantograph windscreen wiper does a really excellent job of keeping the screen clear, and works perfectly well at the higher speeds that can sensibly be reached in wet conditions. This pantograph design has been time tested from the Countach era, although in that car there is a second, much smaller, wiper blade attached to the bottom of the pantograph to keep the lower screen clear, while the Murciélago only has the single large blade. The wiper motor can be set to three different speeds: intermittent, normal speed and high speed. At the intermittent speed setting, the wipe frequency can be further altered to any one of five settings by twisting the wiper stalk. In addition to this, at any given twist setting, the wiper frequency automatically goes onto a higher wipe frequency once a speed of 100km/h is exceeded, as a safety precaution. In a similar protective vein, the high speed wiper setting can only be selected at speeds below 250km/h; a not so subtle hint that one shouldn't be pushing one's luck.

Immediately below the headlight covers is a black section that holds the headlight washer jets. Both the windscreen and the headlight washers are controlled by one of the steering column-mounted levers, and the headlight washers only work when the lights are switched on, but cease to operate once the Murciélago exceeds 130km/h.

The wing mirrors are placed on long support arms, but even so, the wide rear wings are what are principally seen through these mirrors. The wing mirror position can be altered electrically from the driver's seat, and the mirrors can also be electrically folded; both actions are achieved through a switch on the plateau of the central console. Since the nose of the car is narrower than its wide rear flanks, the extended wing mirrors serve as a useful guide as to whether the back of the car will also pass through a given gap -- the overall width of the car is 2058mm with the wing mirrors retracted, and 2240mm with the mirrors extended.

The cockpit rear view mirror is of the non-reflective variety, but only gives a limited straight back rear view, which is further compromised by the Miura-like slats over the engine lid.

THE VARIABLE AIRFLOW COOLING SYSTEM (VACS)

The prodigious power output of the Murciélago engine brings with it the need for a serious cooling system.

The thermal efficiency (a percentage measure of the inherent chemical potential energy of the fuel that a given engine uses, that is fully converted by that engine into the kinetic energy of the vehicle driven by it) of modern production petrol internal combustion engines is notoriously poor, averaging about 25 per cent. The remaining 75 per cent of the energy is converted to heat energy.

This heat needs to be dissipated quickly, and is a challenge, particularly when a large engine is tightly packed within a confined engine bay, and when the overall aesthetics of the body further limit the possibilities for the heated air to escape from the engine compartment.

Liquid cooling is much better than air cooling in solving this heat dissipation challenge, which in turn allows for greater power output with better fuel consumption, as well as reduced emissions from an engine of a given cubic capacity; and underlines why Porsche chose to forego one of the two unique selling points of their classic 911 model-line, in the evolution of the air-cooled 993 variant into the water-cooled 996 variant.

However, even liquid cooling ultimately needs the heat energy to be transferred to the atmosphere, and hence the need for lots of air intakes for input to the radiators, and vents for the heated air to exit.

Large air intakes carry with them two drawbacks. Firstly, they don't look good. As one example, observe the very clean, pure lines of the prototype LP500 Countach, compared to the production cars that carry the less elegant, rectangular air boxes above the mid-mounted radiators. A second example of intakes compromising aesthetics was directly responsible for the birth of Luc Donckerwolke's Murciélago. The Chairman of Volkswagen, Dr Ferdinand Karl Piech, rejected Zagato's proposal for the Diablo replacement because he hated – admittedly amongst other things – the huge ellipsoidal air intakes sitting astride the prototype Canto's rear flanks.

The second drawback of large air intakes sitting in the airstream is that they increase the car's drag coefficient, and so adversely affect its acceleration, top speed, fuel consumption and emissions.

A Murciélago travelling at speed on this planet Earth satisfies the two primary criteria that enables Lord Rayleigh's drag equation to be applied to it: firstly that it has a blunt form factor, and secondly that it is able to cause air turbulence as it passes at speed through the Earth's atmosphere.

Rayleigh showed that when a fluid (in this case air) strikes a body (in this case the Murciélago) with twice the velocity, it is automatically accompanied by twice the mass of airstrikes per second. The drag force (from the Earth's atmosphere) on the Murciélago is therefore (NB: broadly speaking) related to the square of its velocity; ie, the drag force increases by a factor of four when the Murciélago's speed doubles, and increases by a factor of nine when its speed trebles.

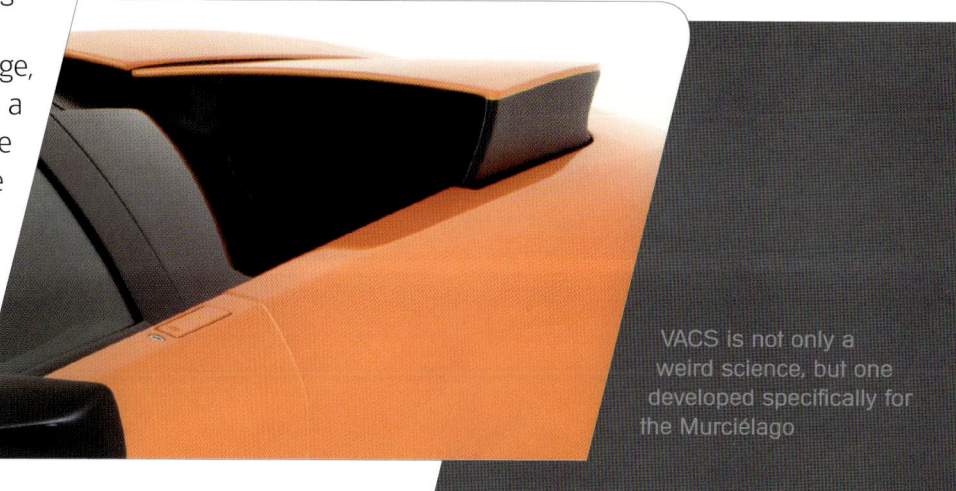

VACS is not only a weird science, but one developed specifically for the Murciélago

So, keeping the air intakes to the minimum size needed to safely and reliably ensure adequate engine cooling, particularly in high ambient temperatures, is important.

At high speed, enough cooling air is usually rammed through relatively small vents. It is at idle and at low speeds – when this ram effect is absent – that the cooling problem is at its most challenging, and larger air intakes are needed.

Lamborghini estimated that in most of its markets, the Murciélago would only need large air intake apertures for about 15 per cent of the car's normal usage. High ambient temperatures in some markets, for example the Middle East, would obviously substantially increase this percentage.

The ideal solution was therefore to have variable aperture air intakes, which could be large when cooling requirements demanded it to be so, and small otherwise.

As beauty is in the eye of the beholder, and some owners might prefer the aggressive look of large intakes, while others would prefer a sleeker look, it would be even better if the positioning of these adjustable intakes could be controlled by the driver from a switch in the cockpit.

It is undeniable, and without shame, to acknowledge that the mid-engined V12 Lamborghinis have always had a major element of theatre about them, and in 2001 the introduction of movable air intakes was a definite crowd pleaser.

Lamborghini had in fact experimented with movable side intake flaps on a test-bed Countach in the late 1980s, but this never reached production. With the Murciélago, Lamborghini designed the variable airflow cooling system (VACS), in which a vent sitting astride each rear flank was smoothly flush with the bodywork in the lowered position, but was able to rise into the air-stream in the raised position. These movable lateral air intakes only have binary positioning: either fully closed (0 degrees), or fully open (20 degrees). With the air intakes raised to the 20 degree position, the intake surface is increased by 80 per cent.

The opening and closing of these intakes are automatically managed by direct current electrical motors, which in turn are controlled by a dedicated ECU (which also controls the movable rear spoiler), depending on the engine coolant temperature, and the ambient air temperature.

The driver can also manually raise the air intakes by pressing a button on the cockpit tunnel console, and should the movement of the intakes be obstructed during either automatic or manual operation, a warning light illuminates on the dashboard instrument panel to alert the driver. The intakes also automatically retract when the car reaches 125mph to improve aerodynamics and to reduce drag.

The drag coefficient of the Murciélago VT is 0.33 with the VACS air intakes lowered, and 0.36 with the intakes raised and the rear wing in its most elevated position.

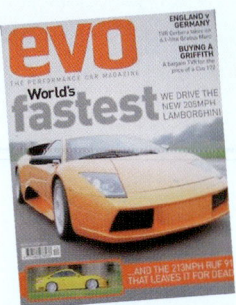

EVO #038, DECEMBER 2001

I can appreciate Lamborghini's desire to let us get our first taste of the Murciélago in a controlled environment but, as David Vivian discovered when he drove the Diablo 6.0 here, Vairano is really too narrow for such a big, heavy car. Despite being carbon fibre-bodied (steel doors and roof excepted) the tubular steel-framed Murciélago weighs a hefty 1650kg. That's a fraction more than the Diablo 6.0, although with its 6.2-litre, 570bhp development of the V12, the Murciélago manages 351bhp per ton to the Diablo's 343. It easily out-muscles the heavier 485bhp Ferrari 550 but has no answer to the Pagani Zonda S, which conjures up 441bhp per ton from 542bhp and a mere 1250kg.

Any one of them would be a handful on this thin, rain-lashed track, which I'm trying to learn as I get to know the Murciélago. There aren't many landmarks but there seems to be plenty of scope for making one – an expensive, crumpled orange landmark just off one of the third-gear sweeps.

Initial impressions are that the Murciélago is easier to handle than the Diablo, thanks to its lighter clutch and the smoother throttle action. The throttle linkage is now drive-by-wire instead of a series of friction-inducing rods and couplings – and that makes traction control a simple addition. Giorgio has advised me that the TCS is 'very firm,' meaning that very little slip is allowed before the ignition and fuelling are knocked back. There's plenty of weight to the steering, which seems to match the general feel of the Murciélago, while the gearshift is typically supercar-notchy until the gearbox oil has warmed through. Even then it's at its most co-operative when you shift positively and double de-clutch on downshifts. Very tactile, very rewarding.

Not that you need to stir the lever much if you don't want to. There's so much low-down urge so cleanly delivered that you can stay in third for the whole lap and still go very quickly. The V12 simply digs in and hauls out from what feels like walking pace without a murmur of protest. For me, this is when the engine sounds at its menacing best, though the heavy beat as it overcomes the body's mass doesn't last long. As soon as the rev counter needle gets to 2000rpm it takes on a nape-prickling urgency and the Lambo thumps forward with a strength that, even in a straight line, makes you grateful for traction control.

As you'd expect after clocking the massive front discs crammed inside those gorgeous alloys, stopping power is phenomenal. It's as hard to activate the anti-lock as it is to trigger TCS on the drying track, and the nose seems to hold up better under heavy braking, as if there's less weight transfer than in the Diablo. If you need reminding that you're at the pointy end of a V12 wedge, a glance in the mirrors, showing those massive scoops raised hungrily into the air stream, does the trick.

Out of the tightest corner I finally succeed in activating TCS. As Giorgio described, it reacts very quickly, catching the tail before it has the chance to swing more than a few degrees, the engine dropping onto what feels and sounds like six cylinders until there's enough grip again. Given that it keeps the tail on such a short leash, there's a surprising amount of low-speed understeer built into the Murciélago's balance. It's impossible to judge the dynamics definitively, of course – the Diablo 6.0 proved much more impressive on real roads, with bumps, cambers and the circumspection that comes with them, than it did here.

Still, later in the day I was able to confirm Giorgio's assertion that the Murciélago is indeed more stable thanks to its engine's lower centre of gravity. It could be sensed through the left-right-left at one end of the track but more so once TCS was disabled and the tail was allowed to edge out under power. You still need to be committed and steady with the throttle but the Murciélago slides and recovers much more cleanly than any Diablo I've driven. A front-engined car like the 550 Maranello still feels less likely to bite your bum if you do have to get off the gas in a hurry, but the fact you can even consider bringing the tail into play around such a narrow track is impressive.

The Murciélago is still very much a Diablo at heart. Easier to handle, more dynamically poised but still a challenge to drive well. It's loaded with character too, a large proportion of which is provided by that glorious V12. The engine's nature is slightly changed so that it delivers more mid-range thrust and a little less top-end vigour, but this will only make it more exploitable and enjoyable on the road.

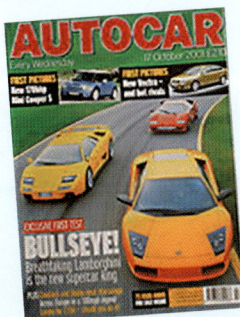

AUTOCAR, 17TH OCTOBER 2001

This time the four-wheel-drive chassis, the new six-speed gearbox and the better balanced anti-lock brakes and steering get equal billing. The Murciélago is, therefore, in chief test driver Giorgio Sanna's words, the most complete Lamborghini so far, as well as the fastest.

Sanna reckons it is so much more composed during cornering and under braking, and just simply faster in a straight line everywhere, that it will lap the old Nürburgring nearly a whole minute faster than the old 6.0-litre VT. The overall lap time, he says, is "maybe a little bit less than eight minutes." For a car that weighs 1650kg this is astonishingly rapid considering a Caterham 500R, usually untouchable around any circuit, managed only 7min 55sec in the hands of a professional.

I can wait no longer, so I turn the key. The explosion of sound behind me is familiar, but more controlled somehow. I blip the throttle and depress the clutch to move away and two things strike me. The accelerator is more responsive and the clutch lighter than of old. Engage first gear and the gearchange is also much more manageable. Already this car feels easier and a lot less intimidating than any other Lamborghini I've driven.

I let the clutch out gently and we start to roll and the ride feels incredibly soothing for a Lambo. As I rumble out onto the circuit over a couple of nasty looking expansion joints the suspension copes beautifully. And I don't have to concentrate to drive it smoothly as you do a Diablo, either.

The six-speed gearbox no longer has a dogleg first to second pattern and the shift is surprisingly effortless considering all the inertia at work down there at the fly-wheel. Into second, then third at no more than 2500rpm, and the V12 feels and sounds much more tractable than I remember. Much more polished.

Into the first chicane and the steering also feels more honest, and sharper. The assistance is quite subtle and it still feels like there's a lot of car behind you when switching from one direction to the next. But there's a precision and a sense of balance present that the Diablo never had. Basically the Murciélago already feels lighter, nimbler and crisper in its reactions – a good 30-40 per cent more so than the Diablo.

Through the second chicane, which is faster, the nose dives towards the apex when I turn in. Yet there isn't a corresponding reaction from the tail when I realise that I've gone in a little bit too quick and brush some speed away with the brakes. The back just stays planted and there's a teeny bit of safe understeer. Right from the word go this car displays a lovely sense of balance.

For the first three laps of this odd little circuit I deliberately don't go full throttle. Because the moment you experience a Lamborghini V12 under load at 7500rpm is the moment you lose all logical judgement about the rest of the car. Before that moment arrives I realize several things about the Murciélago's chassis. Not simply that it rides better, but that it is far more composed than the Diablo's.

In terms of actual grip on offer I doubt there's a huge difference between the two, even though the Murciélago wears tyres that were specifically developed for the car by Pirelli and are, apparently, vastly superior to the Diablo's regular P-Zeros. The engine also sits some 5cm lower in the chassis than before and the springs and dampers have been uprated all round, to improve the dynamics.

The big difference, however, is not how fast you can go through the corners in this car, but how composed it feels while doing so. It's now so agile and immediate in its reactions that, for the first time ever, I'd actually go so far as to call it a sportscar. In the same way I'd call a 360 Modena a sports car. Whereas the Diablo was never anything other than a big old barge which, if you started to throw it around at all, would simply ask you very politely to cut it out. Or just throw you off the road entirely.

But, of course, it's still the engine that ultimately calls the shots. When, finally, my discipline snaps and I plant the accelerator in second gear and hold it on the floor until the limiter arrives at 7800rpm, it's like the solo at the end of an AC/DC concert: everything else that's gone before seems rather irrelevant.

But it's some finale. Lamborghini claims the Murciélago will accelerate to 62mph in 3.8sec and get to 100mph in about 8.0sec. not quite into McLaren F1 territory then, but not far off. It's actually difficult to avoid spouting clichés about what the engine sounds like between 4000 and 5500rpm and then again between 6500 and 7800 when the sensations that come at you are so visceral, so raw. As for the acceleration, well let's just say, it makes the Diablo feel old and slow. Very old and very slow.

Which is astonishing in itself. But the most important thing about the Murciélago, for me, is not that it's so much quicker than its already quite brisk predecessor, it's that fundamentally it hasn't changed in personality one bit. The fact that it's now easier to drive, more friendly on the limit, less thirsty and much more comfortable is great, of course, but the key thing is: it's still the ballsiest supercar on the planet.

In an era where most other car makers are obsessed with protecting us from ourselves, the Murciélago is a gale of fresh air blasting you up both nostrils.

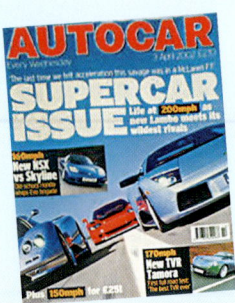

AUTOCAR, 3RD APRIL 2002
That's the thing about the Murciélago: it's the first genuinely user-friendly Lamborghini, yet it manages to retain all the thunderous character of the dinosaur Lambos of old. You still open the scissor doors, free-fall into the cabin and think you have climbed aboard a spaceship. But there's also a depth of engineering nowadays, a logic to the ergonomics that makes you not want to climb out again.

All the controls, including the air-con no less, work, just as they should. The driving position's okay, and at a pinch you can even see out of the back from some angles. It just means there's more opportunity to enjoy the good bits because you spend less time worrying about the bad.

But the focal point, of course, is still the engine, because some things must remain sacred. And once you've experienced what a powerplant like this feels and sounds like up near its red line, the rest of the car tends to fade into the background.

First time the traffic clears properly on the autostrada I buy some space by letting Robinson and snapper Papior get a good half-mile ahead in our VW Bora chase car. In second at 2000rpm the response when I squeeze the throttle is strong, but not heroically so, the relatively modest torque allied to long gearing and high kerb weight preventing the Lambo from leaping forward like a big-engined Caterham.

But at 4500rpm, and then again at 6000rpm, the Murciélago surges into another accelerative dimension as the engine's cam phasing defeats the tall gearing. And by 7500rpm the noise and sheer kick are about as good as it gets.

I do the same in third, fourth and finally fifth and the horizon comes at me in chunks, cars disappearing on the right in a blur as we rocket by at more than twice their velocity.

Someone in a silver Audi wanders out halfway into our lane as we approach at Mach Much and I have to aim the Murciélago through a disgustingly narrow gap between it and the Armco, Harris beside me bracing himself. We make it, at which point I decide I've had enough for the time being. Time to slow down, relax, take in some of this wonderful scenery.

But the Murciélago has already laid down its marker, and continues to do so the rest of the way to Modena thanks to its surprising soothing motorway gait, fine seats and mighty sense of occasion. It really is a refined car in which to eat miles, more so than any previous Lamborghini. The steering's hugely improved and doesn't suffer the kickback or sudden weight up of the Diablo's over ruts and bumps.

Five minutes before we arrive I begin to wonder whether the Pagani and Edonis stand a chance, especially as the Lambo is half the price of the Zonda and a quarter the price of the Edonis, which will cost a cool half-million in sterling when the first customers get their cars later this year.

LAMBORGHINI MURCIÉLAGO ROADSTER

Size and weight
Wheelbase 2665mm (a)
Front track 1635mm (b)
Rear track 1695mm (c)
Overall length 4580mm (d)
Width
– with opened external rear view mirrors 2240mm (e)
– with closed external rear view mirrors 2058mm (f)

Overall height
1132 (with soft top) (f)
1068 (without soft top) (g)
Clearance (without loads)
Front overhang 1005mm (h)
Rear overhang 910mm (i)
Dry weight 1665kg
Number of seats 2

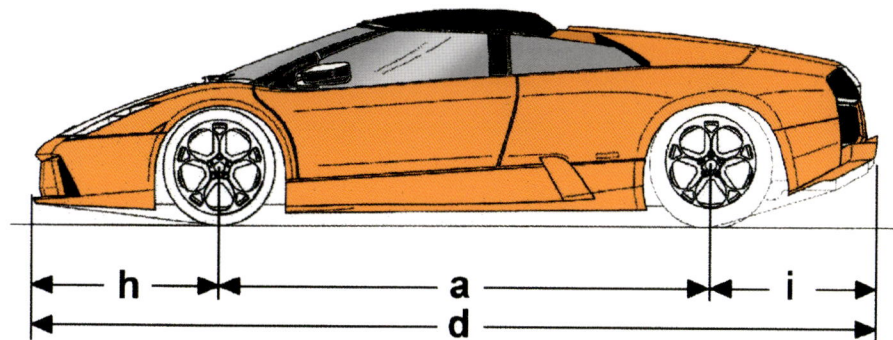

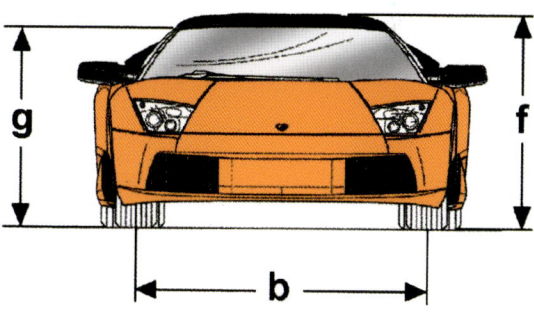

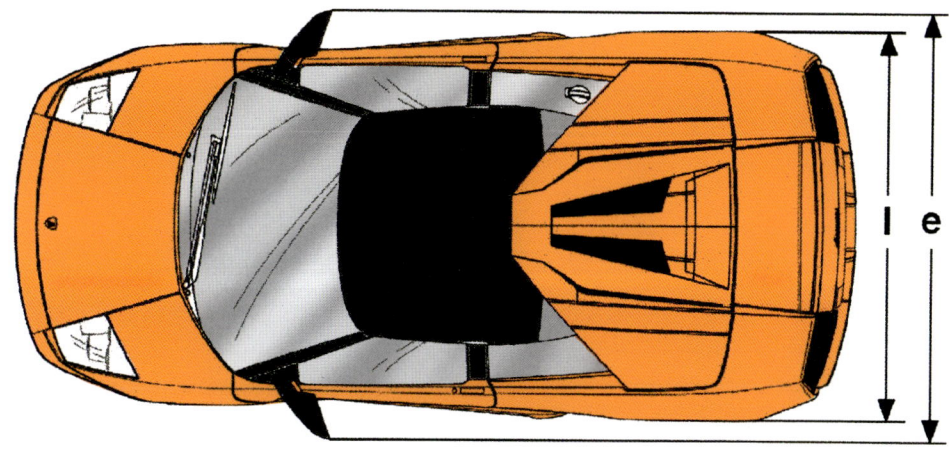

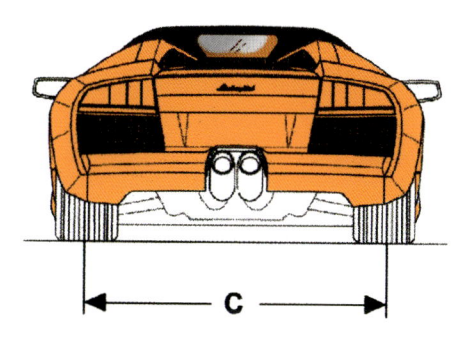

1 Windsheild washer reservoir

2 Steering box

3 Brake/clutch system

4 Gearbox

5 Engine (1)

6 Air conditioning

7 Front differential

8 E-Gear system

9 Cooling system

10 Fuel tank

11 Power steering/ lifting system

12 Engine (2)

13 Rear differential

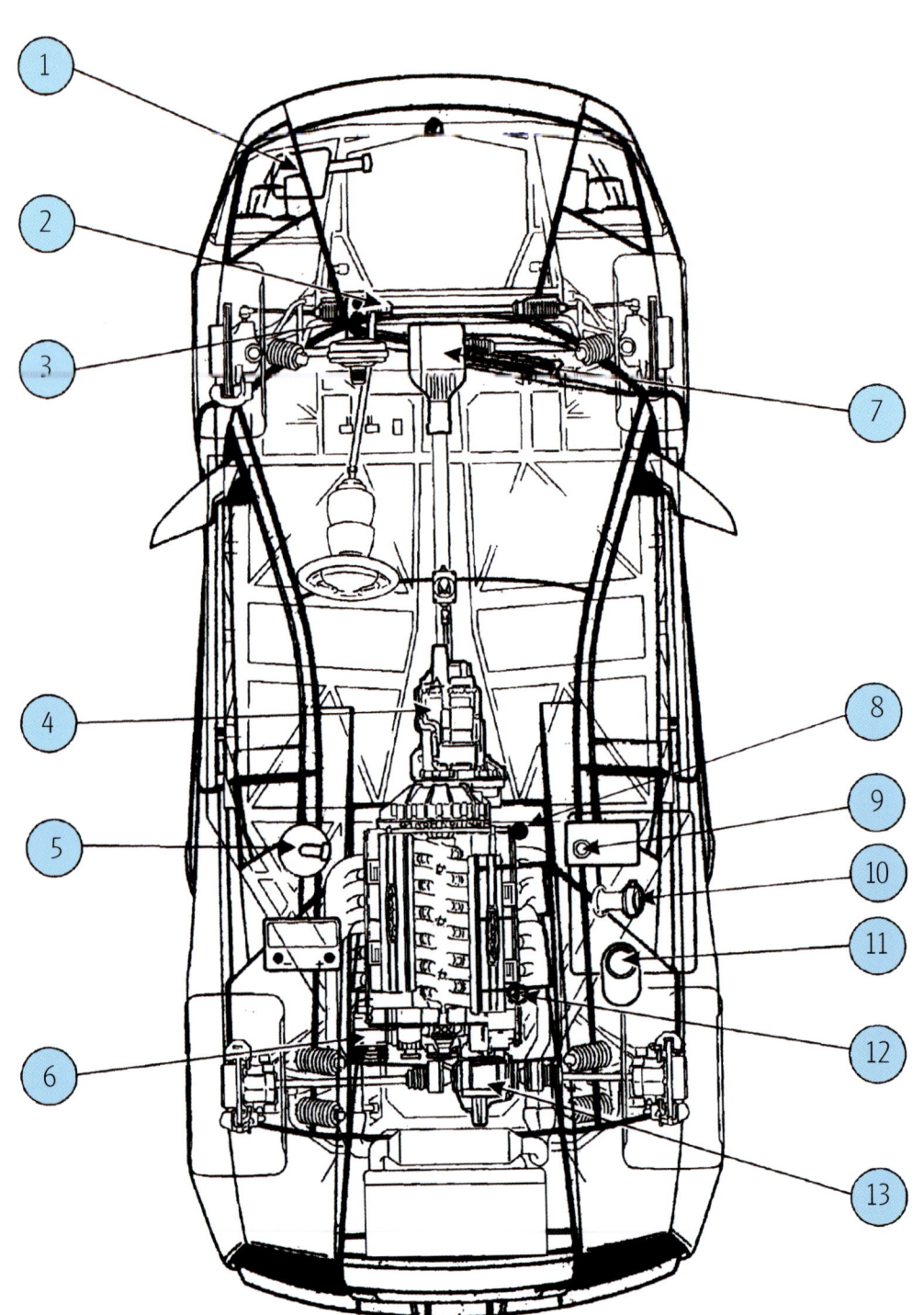

THE MURCIÉLAGO VT 6.2 ROADSTER

*T*he undisputed star of the 2003 Detroit Motor Show was also the new pinnacle of the Lamborghini range; the Murciélago Barchetta concept, the apogee with a detachable toupee.

Mechanically the Roadster VT was very similar to the VT Coupé, with essentially the same drivetrain. The Roadster, however, had modified brakes, softer springs and dampers, and new wheels, but the main story here was the dramatic new styling.

Styled by Luc Donckerwolke in-house at Centro Stile (located at the far left of the factory complex, as you face the front of the factory from Via Modena), the car was simply spectacular. Only the Pagani Zonda Roadster, which also made its debut that same year, could hold a styling candle to it in terms of sensational design, albeit at almost double the price.

Barchetta means 'little boat' in Italian, and, in an automotive context, refers to cars from the 1940s and 1950s, designed or modified for racing. Typically open two-seater sports cars, they eschewed luxury features and styling excess in favour of minimising kerbweight to maximise performance.

The Murciélago Roadster was therefore, to put it mildly, inaccurately named; the car featured new styling, safety and luxury interior upgrades compared to its Coupé sibling, while also suffering from a slightly higher kerbweight and softer suspension settings – both of which would have hurt, rather than enhanced performance. (The production Roadster has a top speed of 199mph, compared to a 205mph top speed for its hardtop counterpart. Their respective dry kerbweights are 1665kg and 1650kg.)

Luc Donckerwolke – the Head of Centro Stile at the time of the Murciélago VT and Roadster's conception, gestation, and birth – openly admits to not liking open-topped cars, because of the inevitable loss of some torsional stiffness, and also because a fixed roof allows for a clean sweeping arc running from the front to the back. This critical styling feature is lost with an open car.

Donckerwolke said that he was inspired by three quite disparate things in designing the Roadster variant of his neo-classic Murciélago VT Coupé: the B-2 Spirit stealth bomber, the 118 Wallypower yacht, and the Ciutat de les Arts i les Ciències in Valencia, Spain.

There is certainly congruity between these three things and the Murciélago Barchetta, in that all are visually jaw-dropping, rare or unique, largely hand-crafted, and very expensive.

The two-crew, nuclear-capable B-2 is a long distance bomber designed to evade anti-aircraft detection, so allowing penetration deep into enemy territory. This flying wing aircraft, with essentially no fuselage or tail, is more properly known as the Northrop Grumman B-2 Spirit, and looks like a sharp-edged 90-degree boomerang with a large glass-fronted central bubble (the cockpit), flanked by two smaller bubbles (the engine air intakes). Only 21 of these aircraft were ever built, with each unit costing on average about $2 billion (in 1997 dollars), once development and testing costs were taken into account. The advanced stealth technology featured four different types of camouflage: visual, acoustic, infrared, and radar, two of which most certainly do not apply to the Murciélago Roadster.

The B-2 has anti-reflective paint and dark undersides, to reduce its visual signature, while the Barchetta concept, even in the most muted body colour, screams out for attention, even more so than its already extrovert hardtop sister.

While the engines of the B-2 are buried deep within the fuselage, and also feature sonic blanketing technology to reduce its acoustic signature, the production Murciélago Roadster featured a sports exhaust as standard in many markets, to better warn and delight on-lookers of its imminent arrival.

There is no readily obtainable data on the infrared or radar signatures of the Roadster, but the nearest parallel to the former might be the

Roadster's CO2 emission figures. The 163-page owner's handbook gives precise fuel consumption data, but no CO2 emissions data. Numerous other sources quote figures of 500g CO2/km for both the Coupé and Roadster versions of the Murciélago VT (and 495g CO2/km for the LP640). The V12 Murciélago, in whichever variant, has, however, consistently featured in the highest echelons of the list of the world's most polluting cars at any given time. Even with the shallower rake of the Roadster's windscreen, my personal suspicion is that the Roadster has a higher drag coefficient (with or without its soft top fitted), and thus a higher CO2 emissions figure, making it more detectable than the Coupé on this basis too.

The B-2 is designed to have a low radar cross-section signature, with its flat surfaces, absorbent paint, and surfaces specifically angled to reflect radar beams somewhere other than back towards its source. None of these were probably high on the design brief of the Murciélago Roadster. I would have liked to have completed the circle, by discussing how the radar signature of the Roadster compared to

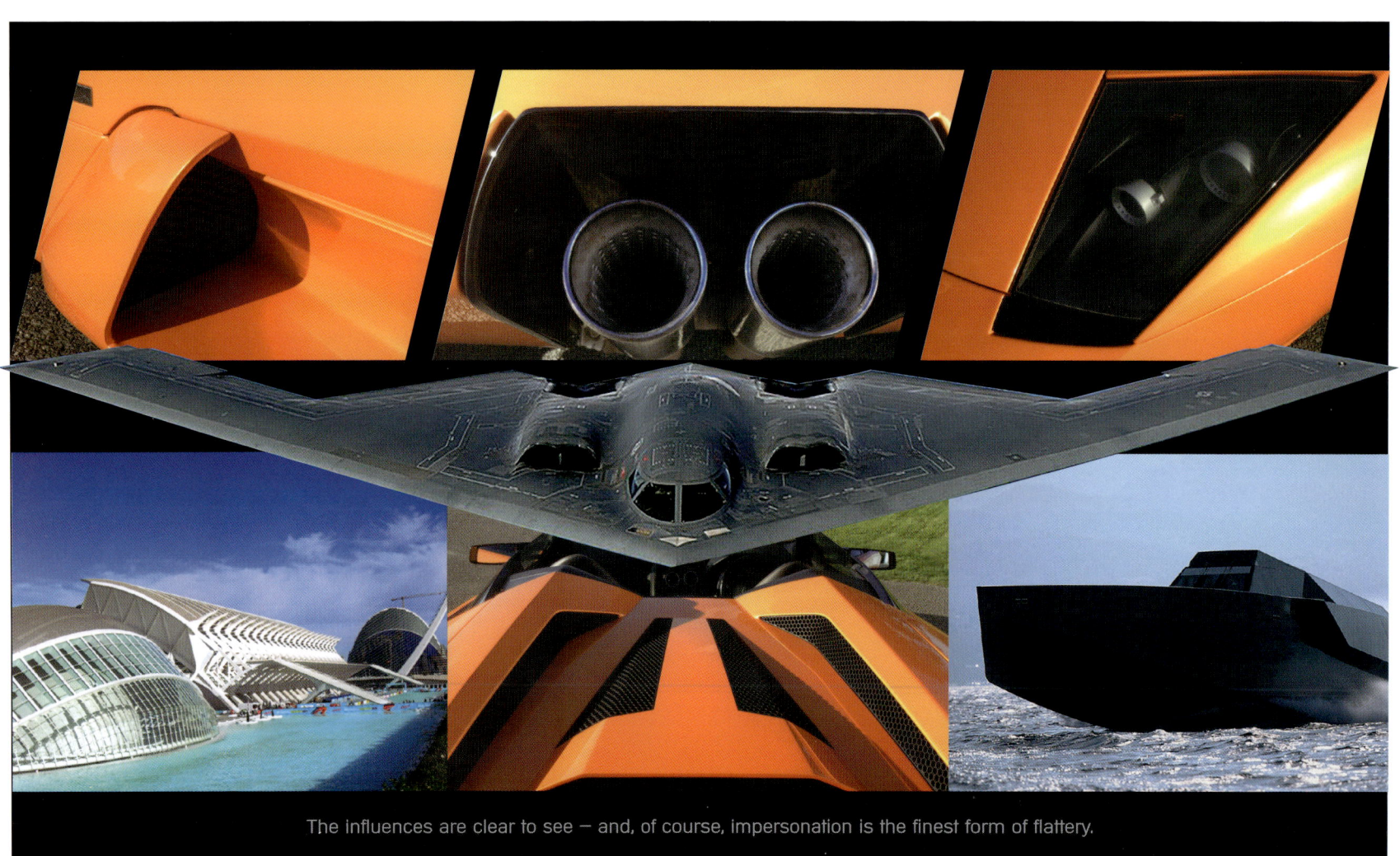

The influences are clear to see — and, of course, impersonation is the finest form of flattery.

the Coupé (the lack of a fixed roof, the flatter, more angled rear engine lid, the metal or optional carbon fibre engine cage). However, even the most superficial scratching of the radar and material science technology papers that I began to study soon put me off pursuing this attempt at drawing a parallel any further; suffice to say that there are many, many variables, including the radar wavelength in question, the wave absorption and reflection characteristics of different materials (carbon fibre versus metal versus glass, etc), and the thickness, the coatings, and the impregnations of said materials. A box that the Pandora in me would do well to leave unopened.

In the boating world, the 118 Wallypower yacht, which was launched in 2003, is just as iconic as any of the guillotine-doored Lamborghini flagships in the car sphere.

In appearance, this superboat's flat surfaces, sharply angled bends with just the odd curve, and black glass housing, most readily recall the Countach, and specifically the one-off LP500 Countach show car.

Built by Wally Yachts, which has manufacturing bases in both Forli and Ancona in Italy (there are some interesting similarities between the lives of Luca Bassani Antivari, the Milan-born founder of Wally Yachts, and Ferruccio Lamborghini – but that is a story for another day), this 118-foot boat has a deep, narrow, 22-degree V-shaped hull. This, together with a straight stem bow, more easily allows it to pierce the waves, and reach its maximum speed of 60 knots. The three Vericor TF50 gas turbines produce a combined maximum of 16,800 horsepower, and so far only one example of this $33 million boat, which can carry six guests and six crew, has been constructed.

Like the body of the Murciélago Roadster, the topside of this yacht, the deck, and the superstructure (to which the glass panes are glued) are very largely made out of carbon fibre. At cruising speed, the 95-ton 118 Wallypower drinks 15 gallons of fuel for each nautical mile that it travels. Unlike a Murciélago at full chat, Bossani says that an outside observer hears only a light whistling hiss as this water wedge passes by at maximum speed: a result of its ground-breaking hull design.

The centrepiece of the Ciutat de les Arts i les Ciències (the City of Arts and Sciences) is the glass-fronted L'Hemisfèric, which has held some professional interest for me since I first came across it in the late 1990s, as it was designed to resemble an eye.

Situated on the outskirts of the city of Valencia, in the south-east of Spain, the Ciutat is composed of the L'Hemisfèric and six other principal buildings, together house a science museum, a planetarium, a laserium, an aquarium, a nature park, a sports venue, exhibition halls, an opera house and an arts centre.

Designed by Spanish architects, Santiago Calatrava and Félix Candela, construction started in 1995, and the L'Hemisfèric was inaugurated in 1998. All seven buildings are quite different, but what

they do appear to have in common are thin lattice-like supporting beams (principally seen in the L'Hemisfèric, the Museu de les Ciències Príncipe Felipe, the L'Umbracle and the El Pont de l'Assut de l'Or), that to my eye, resemble the windscreen and the side screen supports in the Murciélago Barchetta concept and the production Roadster. The buildings also feature large expanses of glass (especially the L'Hemisfèric and the L'Oceanogràfic), that recall the almost wraparound look of the Murciélago Roadster when its windows are raised. While most of the Ciutat buildings feature large radius curves, the details within do show many straight lines and definitely angled bends that mirror the rear engine cover of the Barchetta. Looking at pictures of the L'Agora and the Barchetta side-by-side, the vents of the Roadster's engine cover appear to have been inspired by the L'Agora, though, as the L'Agora was only completed in 2009, maybe it was the Murciélago Roadster that inspired the L'Agora, and not the other way around. Actually, there is great similarity between the vents in the engine lid of the 1985 front-hinged Countach QV, and the vents in the rear-hinged engine lid of the Murciélago Roadster.

A principal argument for the Murciélago Roadster must be that if you are going to indulge in one of the most expensive, most extroverted, most attention-seeking, widest, lowest and least practical cars available, then you might as well do the job properly and with style, and go for the most extreme, and most eye-catching variant of that model.

In the real world, the additional compromises that the Roadster brought with it were: firstly, slightly reduced security due to the detachable fabric top rather than the steel roof of the Coupé; secondly, the effort needed to erect and dissemble the soft top, and thirdly, the fact that the folded and packed soft top filled most of the available boot space.

It is debatable just how often that perfect combination of al fresco Murciélago motoring is served up … the endless blue sky headroom, wind rustling through your hair, easy access to the scent of the meadows, the caress of gentle sunshine on your bare nape and forearms, and the enhanced surround sound V12 audio. Or, whether reality more commonly shows itself as a damp, foggy morning drive to work, having only just squeezed beneath the low, black, unlined fabric roof without dislocating a back joint, followed by being stuck in a traffic jam with the lorry in front belching diesel particulates directly at face level, and studiously trying to ignore the stares of unadulterated hatred and envy from the Greens, that penetrate the exposed cockpit of the Roadster more easily than that of the Coupé. All this before mentioning the melanoma-inducing ultraviolet rays hammering relentlessly on exposed skin, while being hidden at truck axle height with rear view mirrors carrying blind spots the size of galaxies.

Chassis number 1564.

Sky's the limit: The graphic instructions for the Roadster's practical, but perhaps complicated, soft top mean well – but may leave owners wishing there were quicker methods of rain dodging – but once in place, looks as good as it should. Obviously, seconds after you have spent ten minutes or so fitting it, the sun will, no doubt, re-appear.

What is not up for debate is just how wonderful the Roadster looks. At 42 inches tall with the soft top off, and 80.5 inches at its widest point, the Roadster presents a menacing stance tempered with sheer beauty. Slightly lower, it also has a more raked windscreen than the Coupé. With the soft top removed and the side windows raised, the Roadster had an expanse of wraparound glass that resembled the one-piece, curved windscreen of the Lancia Stratos (Tipo 829) that won the World Rally Championship in 1974, '75 and '76. The Stratos featured this curved windscreen so that the Championship-contending driver could still plot his intended course while the car travelled sideways, as it under- and over-steered. With the Roadster, beauty was the reason. Incidentally, Marcello Gandini, who designed the Stratos, also designed the Lamborghini Miura, Countach and Diablo.

The windscreen merges smoothly with the side window glass past the A-pillars, and then gradually slopes downwards to join the bodywork just ahead of the variable airflow cooling system (VACS) vents. In either the elevated or retracted positions, the movable VACS 'batwings,' marry up beautifully with the sloping lines extending from the top of the windscreen down to the engine lid.

The rear engine lid is in itself a work of art, and was specifically designed for the Roadster, so as to integrate with the pop-up roll-over bar mechanism that is only present in the open-topped car.

This rear lid is rear hinged in tribute to the Miura, but more closely resembles the front hinged engine cover of the downdraught carburettor-equipped Countach QV. There is a large central lump, as per the QV, which motoring journalists of the 1980s compared to a gun turret. There are ventilation apertures along the sides of the engine cover to better allow heated air from the engine compartment to escape; one of the main engineering challenges with all the mid-engined V12 Lamborghini flagships being engine cooling management.

As with the Countach, the silver engine lid release lever is hidden within the door jamb, and, as with the Countach, it pays to be very careful that a sudden strong gust of wind does not distort the hinge mechanism, or even rip away the whole engine lid, which can act as a big sail.

The other factor that is undeniably in favour of the Roadster, is that the full orchestra of the Bizzarrini V12 can be better heard in its full glory. The south-north engine is almost in the cockpit of the Coupé anyway, but in the Roadster there is even less material between it and the driver's ear. Admittedly there is more wind and tyre noise to contend with, but at low and medium road speeds the engine music always wins through. This music is lost at very high road speeds, but again in the real world, with an average driver, your concentration needs to be elsewhere, especially as the wind turbulence within the exposed cockpit increasingly becomes yet another distracting factor.

The audio of the Roadster is best experienced by driving from low to high engine speeds at relatively low road speeds, ie, accelerating through the rev-range in the lower gears. The engine sound is a joy even at idle, but once past 5000rpm, and racing towards the red-line, the thundering crescendo becomes mildly frightening and exciting in equal measure for the passenger and on-lookers – who have no control over the loud pedal – and potentially terrifying for owners, in case the roar portends the pistons meeting the valves and attempting to escape through the engine block. Yes, you definitely hear more in the Roadster.

At launch, the Roadster cost about £15,000 more than the equivalent Coupé, with the optional E-Gear transmission system costing another £8000. The mechanicals were largely unchanged, with identical V12 engines, CEEMA six-speed manual or E-Gear robotised gearboxes, and four-wheel drive systems. However, the Roadster has softer spring and damper rates, so it rides better over sharp edges and sends fewer tremors from road imperfections through to the bodyshell.

The Roadster gains an extra 15kg in kerbweight over the Coupé, and a large part of this is due to the extra bodyshell strengthening needed to compensate for the loss of rigidity brought about by decapitating the Murciélago. In particular, Lamborghini further strengthened the floorpan and the door sills, with additional metal and carbon fibre composite bracing.

There is also a new lattice-like engine cage mounted over the engine bay to provide additional rigidity to the rear bodyshell. The standard engine cage is made of black painted steel, but an optional carbon fibre cage could be ordered at a £3000 premium, which brings a 5kg weight saving and adds glamour to the engine bay.

The gorgeous monoblock, five-spoke, 18-inch Hercules wheels made their debut on the Murciélago Roadster. Having more than a passing resemblance to the early Lamborghini five-holed Bravo wheels, and also the later OZ Countach wheels, they hugely enhance the appearance of this open top car. These new wheels also allow additional space for larger, upgraded brakes.

The original brakes on the VT Coupé, while more than adequate for normal road use, were found wanting with high speed road use and track use. The Roadster therefore featured new brakes with larger diameter and thicker dual compound discs, now 14.96 inches in diameter at the front, and 14.0 inches at the rear – in fact, exactly the same size as on its cousin, the Audi Le Mans Quattro Concept (and up from the 14.0 and 13.2 inches, respectively, found on the Murciélago VT Coupé). New eight-piston front callipers and an even more powerful revised brake booster, to reduce brake pedal effort, were introduced at the same time. With these improvements, the Roadster could decelerate from 125mph to standstill in less than 130 yards, and brake fade from repeated high speed decelerations was also markedly reduced.

The fabric soft top is undoubtedly difficult to fit and remove, mainly because the main component is large, heavy and unwieldy, with protruding side flaps, and contains within it a small but vulnerable glass window.

While it is possible for one person to fit the soft top, this involves taking an unacceptably high risk of scratching the three-layer paintwork. Ideally, the soft top should be erected by two people who have taken the time to read and absorb the instruction manual, with no distracting on-lookers present.

The owners handbook warns that water might enter the cockpit through the front crossbar, or the front and rear joints of the lateral gaskets, but this has not happened to me even when making "sensible progress" in heavy rain.

The soft top is composed of four major components. Firstly, the main fabric body, carrying within it the glass rear window, rubber gaskets, metal securing pins, Velcro and six press studs, as well as two lateral rear Tenax fasteners. The remaining three components are a right and left upright, and a removable central crossmember.

First, the large sail-like engine compartment lid has to be raised (which can be a real issue, as wind so often accompanies rain, and having a person hold the raised engine lid steady against the wind can prevent an expensive accident), as do the two small rotating pressure-fit panels behind the seats. The right and left soft top securing pins then have to be carefully inserted into their respective holes, which were previously hidden beneath these pressure-fit panels. This requires restrained force, but once the pins are securely locked into position, the weight of the main body of the soft top is held by these two pins, which means that the rest of the fitting procedure becomes much easier.

The soft top is then partially opened by pushing its leading edge towards the windscreen, at which point six rotating support arms (three on each side) are released from the frame, and temporarily come to rest in mid-air. The soft top is now pulled all the way forwards onto the top of the windscreen surround, while achieving correct alignment by lining up the centring pins with their respective holes. The right and left uprights are then fitted between the windscreen and the rear bodywork, and the anterior two free-floating arms are laid to rest on the uprights, while the posterior rotating arms rest on the rear body. Next comes what I find to be the most difficult part of the whole process, which is the insertion of the small plastic wedge found at the very front of the soft top into a gap at the top of the windscreen. My persistent fear is of tearing this fragile looking plastic wedge. Sourcing a replacement wedge, and finding someone with

the requisite skill to reattach the new replacement wedge, is the stuff of nightmares.

Next, the removable central crossmember is fitted to the soft top frame, and then it is the simple matter of pressing down on the Velcro between the fabric soft top and the uprights, attaching the six rear button fasteners, and finally securing the two rear lateral Tenax fasteners. Simple really. The engine lid can now be closed with a sigh of relief, and the soaked leather interior wiped down with a towel, if one did not have the foresight to start the process well before the heavens opened.

The disassembly process is essentially the reverse of the procedure outlined above. The very first time I fitted the top, I did it myself having studied the manual carefully beforehand. It took me 45 minutes, and I only just managed to avoid scratching the bodywork with the securing pins. For the subsequent disassembly, I had the sense to secure the assistance of my wife, but even so, it took us half an hour. We can now do either the fitting or the unfitting in about ten minutes. Once fitted the soft top is impressively wind and watertight – much more so than the owner's handbook would lead you to believe.

Lamborghini states that "the soft top was designed as emergency protection in case of rain," and warns that the soft top is not guaranteed to stay attached to the car at speeds beyond 160km/h (99mph). Many articles suggest that this is a major limitation of the Roadster, but, in so very many countries, maintaining a speed over the soft top limit for any significant period of time would result in the loss of one's driving licence at best, and the loss of one's liberty at the other end of the spectrum.

Much more of a real world consideration is just how much more quickly the Murciélago consumes fuel during sustained high speed driving, which, in turn, can severely limit the touring range of the car. An interesting – if only *just* relevant – aside to the Roadster's fuel consumption, and therefore its real world practicality, is how the movement of the soft top with increasing speed affects the car's drag, and so its fuel consumption. At standstill, the maximum height of the car without the soft top is 1068mm, and with the soft top attached is 1132mm. Bernoulli's principle explains why the flexible fabric soft top rises further (admittedly slightly, due to the metal frame) into the airstream at speed, which in turn leads to poorer fuel consumption.

Daniel Bernoulli (1700-1782) was born in Holland into an acclaimed family of mathematicians, and first studied mathematics, and then medicine, in Venice, before settling down in Switzerland. In 1738 he published *Hydrodynamica*, a study of how fluids behave in motion, which asserted that as a fluid moves faster, it produces less pressure. This principle explains how lift is generated on an aircraft's wings, why fluid flows through a carburettor, and also why the soft top of the Murciélago Roadster rises as the car's speed increases.

Two views of the Randy McPherson tonneau. Water resistant and fitted within about 90 seconds. Both time and leather saving, thanks to the unique Path-designed modifications!

The drag coefficient (Cd), is a dimensionless quantity that is used to quantify the resistance of an object in a fluid environment (its drag), and is always associated with a particular surface area. As the Murciélago's soft top rises with increasing speed, it presents a slightly increased surface area (the reference area A), to the airstream, and as the drag force on the Roadster has a linear association with both the Cd and the reference area A (drag equation: $Fd = \frac{1}{2} p\, u^2\, Cd\, A$, where Fd is the drag force), you can see that the faster the Roadster goes, the greater will be the lift on the soft top, and hence the reference area A, and so the greater will be the overall drag on the car.

The overall drag on the Roadster will, however, be much greater without the soft top on, due to increased air turbulence, particularly around and within the open cockpit

So, for those Murciélago Roadster owners keen to burnish their green credentials, keep the soft top on, as the overall drag on the Roadster will be less, and also keep well under the 160km/h soft top stipulated maximum.

A BESPOKE TONNEAU

With practise, the factory soft top can be erected in about ten minutes. However, ten minutes is eight minutes too long when you live on a notoriously wet and windy island. A cold and wet owner/driver can easily be warmed up and dried out by the car's heating and air-conditioning systems, and so barely deserves mention. Partners need more respect, but, most of all, acid rain spots disfiguring the leather interior need to be carefully avoided. Additionally, while the carefully folded and packed factory soft top fits snugly into the front luggage compartment, in doing so, it takes up so much of the available space that there is little space left for touring luggage.

I therefore started to search for a soft top that could be fitted quickly, and that would take up much less of the available luggage space. Together with Mr Google, I stumbled across Shamrock Trim in Florida, which already had a template for what, in the 1940s and 1950s, would have been a classic shower cap for a barchetta: lightweight and minimalistic, while also meeting my requirements of being quick and easy to fit, and compact. However, I also wanted the tonneau to be able to be used while driving at low speeds; something that would not necessarily pass national legislation requirements, but could be used on private land, for example moving the car from the driveway into the garage when confronted with a sudden downpour, or when leaving the car at a show, where the presence of a soft top would be both a security measure, as well as protection against unexpected inclement weather. I therefore suggested three modifications to Shamrock: a rain strip along the tonneau-windscreen border, additional and stronger attachments to secure the tonneau to the door mirror stems, and, finally, a Velcro-based securing device to hold the loose rear tonneau flap tight against the adjacent bodywork. Randy McPherson of Shamrock Trim was kind enough to indulge me with these modifications, and I am delighted with the result. This tonneau is made of a single layer of waterproof canvas and three metal bars, and requires no modification to the car at all, as all the fastenings attach to the factory studs already present. It takes all of about 90 seconds to take on and off, and occupies only around ten per cent of the luggage compartment space. A much recommended accessory for the Roadster – ask Randy for the Path modifications.

WIND-SCHOTT AND TOP SPOILER

With the soft top removed, the Roadster has two optional factory accessories to reduce in-cockpit turbulence.

The first is what the factory intriguingly calls the 'wind-schott.' This is essentially a detachable glass pane with a black surround that incorporates two locating pins. These locating pins fit into the same holes that secure the soft top, and are located beneath rotating flaps, behind the seats. The wind-schott comes with its own black travelling bag emblazoned with the Lamborghini logo.

The second item is a spoiler that attaches to the top of the windscreen surround. It diverts the airflow away from the cockpit, and so improves driver and passenger comfort. This spoiler comes in two separate mirror-image pieces, so that, when separated, each piece fits into the front luggage compartment.

The spoiler is assembled by aligning the protruding pin on one of the pieces with a corresponding hole in the second piece, and pushing the two pieces together. Once assembled, the spoiler is secured to the windscreen surround through a rotating clasp mechanism.

The aesthetics of the Roadster are somewhat compromised by the use of either of these options, but each genuinely improves long distance, open top driving comfort.

ACTIVE ROLL-OVER PROTECTION SYSTEM

In the absence of a solid fixed roof, the Murciélago Roadster has a hidden active roll-over protection system (AROPS), to meet the conflicting demands of maintaining the clean, elegant lines of the Roadster, while also protecting the driver and passenger in the event that the car becomes inverted. To accommodate the bulky AROPS, the engine lid of the Roadster had to be completely redesigned, and had to become rear hinged. The AROPS, supposedly adapted from the Audi A4 cabriolet, is made up of two roll-over bars (one behind each seat), roll-over sensors distributed throughout the car, an electronic control unit, and seatbelts with pretensioners.

The AROPS works equally well with or without the soft top attached.

The roll over sensors send signals to the electronic control unit, describing the degree of vehicle tilt, and, once a set threshold is reached, the electronic control unit activates the roll-over bars into their raised position within milliseconds, and then activates the seatbelt pretensioners. Both these protective mechanisms are also automatically activated during head-on, lateral and rear-end accidents, depending on the degree of induced tilt.

When the car's ignition switch is turned on to its first stop, a warning light showing a red car tilted onto two wheels lights up on the dashboard display, and an automatic check of the AROPS takes place. If a fault is detected, the red warning light remains illuminated, but in the absence of any faults the light turns off, indicating that the AROPS is fully functional.

THE ASYMMETRIC INTERIOR

One notable interior styling feature of the Roadster is that the driver and passenger sides of the cockpit are styled with different types of leather. On the driver's side, the seat and the door panel are all clothed in perforated leather, while on the passenger's side, the corresponding parts are covered in smooth leather as standard. The car also has a non-reflective rear view mirror which is particularly useful when driving in sunny conditions with the soft top off.

MOVIES

The Murciélago Roadster has featured in several movies, including the following two, small clips of which can be seen on YouTube.

In the film *Transporter 2*, which was directed by Louis Leterrier and released in 2005, a gleaming black Murciélago Roadster makes the point of just how low it is, by driving beneath the traffic barriers at the airport, and then further emphasises this point by passing first underneath the wing, and then underneath the fuselage, of the villain's executive jet plane.

Batman Begins, produced by Warner Brothers and directed by Christopher Nolan, shows billionaire Bruce Wayne (Batman) – played by Christian Bale – driving the Murciélago Roadster accompanied by two exquisite ladies, who, despite their short attire, are both able to exit from beneath the raised passenger door of the Roadster most elegantly. The car has particular poignancy in this film, as Murciélago means 'bat' in Spanish, and, with the movable side air intakes in their raised position, the car does resemble Batman in his classic stance, with his cape held wide open.

Top: Aside from minor changes, the original Murciélago Roadster concept was almost correct from the word go.

Left: 'AWE55OM' carries the 'Hercules' wheels which first featured on the 6.2 Roadster. They are designed with heat dissipation in mind, so are practical as well as beautiful

EVO #071, SEPTEMBER 2004

What better way for captains of industry to blow the cobwebs out of their stressed heads than kicking the Roadster's colossal V12 into life? If you want an engine that shakes you to your very bootstraps, this is it. Angry, intense, verging on violent, it starts with a boom and idles as though Lamborghini has forged the forces of nature into the engine block.

We're in the E-Gear car, so pulling away involves nothing more than depressing the brake pedal, pulling back on the right-hand paddle and gently easing in the power. It's a smooth enough process, but you can sense the electronics are struggling to balance the clutch against the mechanical drag of the all-wheel-drive transmission, particularly when you've got a turn of lock applied.

Once up and running, the e-gear 'box is happier, although it does encourage you to aimlessly doodle up and down the gearbox, rather than pick a gear and stick to it, as an engine with such mountainous reserves of power and torque allows you to do. Perhaps discipline comes with familiarity, but there's something uniquely satisfying about short-shifting from second to fourth to sixth that you simply can't do with E-Gear.

It's a scorching day, and the Italian sun is enjoying roasting some colour into our lily-white forearms and foreheads through the open roof. We'll be burned for sure, the combined effects of the air-conditioning and a surprisingly gentle breeze swirling around the cockpit lulling us into a false sense of security.

Until now it's been too busy to give the Murciélago its head, but with the traffic thinning and the roads becoming more challenging, the time has come to introduce the long travel throttle pedal to the carpet. Most cars need a moment or two to gird their loins, but the Lamborghini shoves forwards instantaneously, the force increasing crazily with every 500rpm gained. It's an incredible sensation and one that is rarely interrupted by gearshifts, thanks to a stride of Sharapovian proportions.

We're on a twisty but by no means tortuous hill road, but such is the reach of second gear that we rarely need to grab third, the Roadster devouring whole stretches of road like a kid sucks up spaghetti. When you do get the chance to take a higher gear there's the same insistent, incessant tide of acceleration that places you entirely at the mercy of the forces of g. Few engines allow you to flex between walking pace and warp speed quite like the Lambo V12.

That the Murciélago Roadster is dizzyingly rapid isn't a great surprise. What's shocking is that it corners with undiminished zeal, combining massive grip levels with the poise and adjustability that made the Murciélago Coupé such a step change over the outgoing Diablo 6.0. It has progression and predictability in spades, the nose gently washing wide as you reach its adhesive limit. Back-off and the nose dives back on track, with no fear of provoking that pendulous V12 engine into sliding past your left or right shoulder.

Leave the electronic safety nets engaged and you can make extremely rapid progress, but under full-bore acceleration in the first three gears, crests and bumps can momentarily trigger the ASR, which causes the fluently accelerative V12 to stammer, even though in truth the monstrously wide tyres have everything very much under control.

In the dry, at least, you have no cause to fear the Roadster, so why not switch the ASR off and treat yourself to the full, uninterrupted, uncensored, widescreen, Dolby 5.1 Lamborghini experience? The noise, the sense-scrambling acceleration, neck-straining cornering and now, thanks to those huge brakes, eye-watering deceleration, are all there. Only now, thanks to your exposure to the elements, the experience is so much more intense.

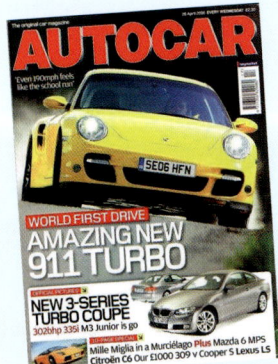

AUTOCAR, 26TH APRIL 2006

A Murciélago would look dramatic even if it were grey on a grey day, but in orange under a brilliant blue sky it highjacks your attention like a firecracker at a funeral. The black hood and the graphite-grey wheels make it look meaner than an angry bouncer, and if you still feel its presence falls short on the malevolent front, you can electrically raise the airscoops flanking the engine. Normally they raise automatically, to shovel the cooling air into the maw of a hard-worked and hot 572bhp V12.

All of which makes you want to get in it. Press a finger on the front edge of the door handle to flick it free, grasp it, heave and the big door cantilevers skywards. From there it's a short, crouching drop into the seat. A slightly offset wheel, a full set of surprisingly sober instruments and an absolute constellation of warning lights – including an alarming one depicting the Murciélago at a 45-degree tilt to warn of rollover bar deployment failure – confront you in a cockpit that is beautifully crafted. You sit amid a mix of unwavering double-stitched leather – you wouldn't have seen that in Lamborghini's past - whose contrasting orange threads look terrific. It feels rich and workmanlike.

And of course, you will be wanting to remove the roof, for which you will need a good 15 minutes. You'll be plucking at studs, peeling Velcro, unhooking rods and carbon spars and compressing them, folding, manoeuvring, tugging, jerking and – if things go badly – swearing and ripping knuckles. This roof isn't easy to remove, and re-assembling it is a memory test as much as a manual work-out. You'll master it with regular use, but beware cloudbursts.

Even if you've driven a Lamborghini before – and I have had the privilege several times – it's impossible not to feel intimidated by its bulk and the feeling that it will be difficult to see out of. In fact, you can see ahead with ease, and with a paddle-shift transmission, driving it is go-kart simple – so there's no embarrassing stall as we leave the factory gates.

Apart from your proximity to the ground, what hits hardest is the noise. It doesn't take long to realise that the gnashings of the V12 are coming at us through a roof whose fabric is close to useless as a sound-deadener. It's a noise potent with promise, but it does undermine an otherwise comfortable car. You can pretty much get the seat where you want it – virtually impossible in the old Diablo – the major controls are to hand, the minor switches easy to pinion with a digit. The ride is surprisingly pliant too. The car follows the contours of the road pretty closely, but you're rarely jostled to distraction. So if the journey to Brescia is slightly frustrating – the only chance to even begin exercising this car is away from the autostrada tollbooths, when the unfettered bellow of the V12 tempts you into a lunge at oblivion – it's still a real pleasure. A pleasure for others on the road too, who regularly photograph it with their phones.

Such troubling thoughts will be banished when you reach one of the

... Apart from your proximity to the ground, what hits hardest is the noise ...

highlights of the Mille Miglia drive. There is scenery here of the verdant, mountainous kind, and a route that rises and falls as much as it twists. Peel off the roof and you are going to feel more a part of it, intensifying the full physical experience of this car, though it is not actually particularly physical to drive. The biggest effort will be of concentration – this car is not small, and demands precision of the driver, especially on these tight roads. There are big forces to control. After a half-hour, high-speed burst along the Futa pass I feel as if I've had a light – and exhilarating – work-out. Mille Miglia combatants would have been keeping this up for hours at a stretch, and in machinery of wayward character.

But the task is eased by the Murciélago's superb steering, a precision tool bundled with the kind of energised feel that is so rare in modern cars, and a set of carbon fibre stoppers whose resilience, as we plunge towards valley floors through endless confections of twists, is mighty. Actually, they're overlight at low speeds – just like an Audi's, oddly – but that disappears when the going's rapid.

Though this is quite a big car, one of the most astonishing feats is not to feel it when you fling it at second- and third-gear hairpins, and it's impossible to resist the enjoyment of feeling this big brute of a car muscle its way through a bend as if it were a ballerina.

Of course, the Murciélago understeers somewhat if it's launched hard into tight turns. But you can kick the tail around slightly even with the traction control on – an arrangement it's sensible to maintain when your run-off area is a rock-face or a rapid descent onto the switchback road 10m below. It's hard not to admire the bravery – recklessness even – of those Mille Miglia combatants. It was the failure of a tyre, indeed, that killed this race and stole the lives of Alfonso de Portago, who was driving, ten spectators, and his co-driver. That was in 1957, two years after the 1955 Le Mans crash that killed 55 people. Three days later, the government decreed the end of the Mille Miglia and all road races in Italy.

But that ban could hardly kill the pleasure of driving an athletic car hard along a picturesque road, nor the pleasure of seeing spectacular machines streaming through dramatic scenery. It's why beautiful, fast and physical cars, cars like Lamborghinis, are built and bought today. The Murciélago Roadster, though able on a scale beyond the wildest imaginings of a 1930s racer, captures the same spirit, in a small part because it supplies similar inconveniences with its soft roof. The Mille Miglia was an A to A event – it ran from Brescia to Brescia – and the Roadster will likely be used as an A to A car, a car that, like this race, is all about glorious indulgence.

THE MURCIÉLAGO
LP640 COUPÉ

*T*he launch of the Lamborghini Murciélago LP640 Coupé, at the Geneva Motor Show in March 2006, was entirely in keeping with Lamborghini's history of slowly evolving its mid-engined V12 flagships. Witness the development of the Countach from the LP400, by way of various mechanical and styling guises, through to the Anniversary, and the similar progression of the Diablo VT through to the 6.0.

The end of the Murciélago VT's five-year tenure was hastened by three different things. Firstly, established rivals were bringing out almost comparably priced cars – such as the Ferrari 599, and limited edition motor sport variants of the Porsche 911 – that seriously challenged the performance of the Murciélago VT. Secondly, a new group of hugely expensive hyper-exotic cars like the Pagani Zonda, Koenigsegg CCX, and Bugatti Veyron were moving the performance envelope so far forwards that the Murciélago's performance also had to be improved, just to stay relevant within its lesser supercar segment. Finally, there was now an in-house challenge to the Murciélago, in the shape of the Lamborghini Gallardo.

The Gallardo made its debut in 2003, and, in featuring a longitudinally mid-mounted V10, four-wheel drive, stunning sharply creased bodywork, a luxurious interior, and the status of the Lamborghini badge, it was superficially a big and direct threat to the Murciélago.

The Gallardo was also more user friendly in the real world; it had a smaller footprint, was cheaper to buy and service, and was more fuel efficient as well as less polluting.

Lamborghini in turn differentiated and protected the Murciélago from the younger and cheaper, 500bhp 376lb/ft Gallardo pretender by reserving the spectacular up-swinging guillotine doors solely for its flagship V12 range.

In actual fact, this appearance of a threat was a very superficial challenge, in that the Murciélago was a very much more bespoke car than the Gallardo: a Gieves and Hawkes to an Armani, or, maybe more fittingly, a pair of Lobbs to a pair of Churchs. There was a completely different level of hand-craftsmanship and labour intensity that went into building a Murciélago compared to a Gallardo, be it in the internal components of the Murciélago's engine, the tubular steel spaceframe chassis, the more intricate hand-stitching of the leather, or the much more time-consuming, off-site painting of the Murciélago's body panels.

However, with the Gallardo only costing about two-thirds of the price of the Murciélago VT, the latter needed both a styling update and improved performance to justify its price-tag and status at the top of the Lamborghini food-chain. This materialised in the form of the LP640.

In naming this new variant of the Murciélago the LP640, Lamborghini reverted back to the alphanumeric nomenclature of the Countach era.

Decades of Lamborghini V12s, represented in colour from the 1980s to 2000s. Left-to-right: Countach, Diablo, Murciélago (note: Q-Citura interior) and Aventador. In the background, on the water: A fair swap for the lot plus cash either way!

The LP stands for 'longitudinale posteriore' in which the 'L' refers to the fact that the V12 engine is orientated longitudinally within the spaceframe chassis, and the 'P' indicates that the engine is rear mid-mounted. The '640' refers to the 640hp (631bhp) put out by the newly modified engine.

THE BODYWORK

While the LP640 featured many mechanical and bodywork changes over the Murciélago VT, in the main these two cars were very similar, and the following narrative should be read in the context of the 6.2 VT section, as the text that follows only highlights the differences between these two variants, without repeating the many similarities.

Lamborghinis are a visual treat, but after about 2000 VTs, a yet more dramatic body was thought necessary. Undeniably part of this was for marketing reasons, but Lamborghini also had to make body modifications to accommodate the requirements of the new, larger engine, which needed more cooling, and took this as an opportunity to make the car more aerodynamic, to thereby increase performance and decrease emissions.

Amongst the most noticeable changes were the new front and rear bumpers, and the now asymmetrical sill air-intakes, low down in the flanks.

The new front bumper had a more angular appearance in profile, with a more prominent front spoiler, and the front air scoops were now larger and five-sided, rather than rectangular as in the VT. The central part of the lower front spoiler featured an indent, which gave the frontal aspect of the car a more aggressive appearance. This look was further enhanced by the lateral aspect of the air scoop apertures being drawn deep into the sides of the bumpers.

The rear bumper now had triangular downward extensions at each lower extremity, and the central exhaust aperture surround was larger, and again had downward extensions that more completely enveloped the exhaust. Rather than the twin circular exhaust pipes of the VT, the LP640 featured a single large ellipsoidal-like exhaust, which was integrated into the new rear diffuser.

Another significant change at the rear of the car was the new single colour tail lights. These red lights featured LEDs rather than conventional incandescent bulbs, and therefore lit up more quickly when the brakes were applied.

This mid-life make-over of the Murciélago also involved the unusual design of the car not being exactly symmetrical from right to left. In an over-exaggeration of form following function, the right lower air vent at the rear sill was very similar to that on the VT, with its back end pretty much closed. On the left side however, this lower air vent was now larger, more prominent and fully open at the back to maximise airflow. The reason for this asymmetric design lay in the fact that the LP640's 6.5-litre engine system included a single large oil cooler, located on the left side, and the left lower vent therefore needed a large volume of cool air. The right lower vent did not carry this same responsibility, and so could be partially closed off.

The windscreen wiper system and the door-mounted rear view mirrors were redesigned to better integrate into the shape of the LP640, and also to improve the car's aerodynamic profile. The VT mirrors were supported by twin pillars, while those of the LP640 had a single black pillar extending back to the bodywork.

New aluminium Hermera wheels, 8.5x18 at the front and 13x18 at the rear, were now standard. More elaborate in design than the Hercules wheels, and having more than a passing resemblance to the five-spoke Bravo wheels seen on the Countach, these were available in either a silver or a dark titanium finish.

The standard tyres on the LP640 were Pirelli P Zero Rossos with asymmetric unidirectional treads, measuring 245/35 ZR18 at the front, and 335/30 ZR18 at the rear. Optional track/race Pirelli P Zero Corsa tyres were also available, as were Pirelli P Zero Sottozero for use in very cold conditions.

Within the Hermera wheels lay self-ventilating 380x34mm front

and 355x32mm rear steel discs, which were integrated into a dual hydraulic braking system that included a vacuum brake booster, four-channel ABS, electronic brake control and traction control.

An expensive option (costing £7780 in 2006, when the on-the-road price of a basic LP640 without any options was £190,000) was the inclusion of ceramic brake discs measuring 380x36mm front and rear, with six-piston Brembo callipers. This ceramic brake option brought with it negligible brake fade during hard use, significantly reduced unsprung weight with its attendant benefits in handling and comfort, and much longer disc lifespan. Replacement ceramic discs are, however, horrendously expensive, and much more easily damaged, particularly by careless technicians when removing and replacing wheels.

Another expensive option, at £4312, was a transparent engine cover, replacing the Miura-like slats of the engine lid, and allowing an unobstructed view of the 6.5 litre unit. This could be a mixed blessing, as road dust and dirt readily settles on the engine surface when driving in a country setting, and a fastidious owner is therefore constantly reminded of his or her engine bay cleaning duties.

THE MECHANICALS

At the heart of the LP640 face-lift was its modified engine. At the time of the VT's development, it was thought that the Bizzarrini engine could not realistically be stretched beyond 6.2 litres, and that a completely new engine would be needed thereafter. Within Lamborghini's engine department there were two schools of thought: one that believed the above, and one that felt that a further stretch was just possible.

Primary objectives for the modified engine in the LP640 were to increase power and torque outputs, while simwhich was the main criticism levelled at the 6.2-litre unit), and improving mechanical refinement and drivability throughout the rev range.

To this end, the all-aluminium alloy, Bizzarrini-derived unit had both its bore and stroke increased, to 88mm and 86.8mm respectively, to give a total capacity of 6496cc. Lamborghini's powertrain engineers extensively reworked the engine, with a revised continuous variable timing system on both the intake and the exhaust valves, a completely new cylinder head design, a new intake system, a new cooling system, the much larger oil radiator already mentioned above, a new crankshaft, revised camshafts, and a modified exhaust system.

The 6.5 unit was considered a great success, in that it met all the objectives set for it. It was more refined than the 6.2 unit, and it produced more power and torque: 640hp (631bhp) at 8000rpm and 486lb-ft (660Nm) at 6000rpm. The engine in the LP640 had its red line at 8100rpm: 300rpm higher than in the VT engine.

The LP640's six-speed manual gearbox was modified with tougher internals to cope with this increased torque and power, and a stronger rear differential and stronger axle shafts were also introduced.

The single dry plate clutch had a disc diameter of 272mm, and the revised gear ratios in the LP640 were as below:

First gear	3.091:1
Second gear	2.105:1
Third gear	1.565:1
Fourth gear	1.241:1
Fifth gear	1.065:1
Sixth gear	0.939.1
Reverse gear	2.692.1

The E-Gear ratios for the LP640 were also as above.

Lamborghini's 'Verde' colours are almost as startling as its 'Arancio' range. Both colour ways hark back to the days of the Miura, and the most popular ones still create a visual scandal on today's range of Lamborghini cars. Above: Verde Ithaca never fails to look fabulous indoors or out on the LP640.

The E-Gear system in the LP640 featured revised software, which improved gear engagement speed, and also improved clutch longevity. E-Gear cars could also be equipped with Lamborghini's Thrust Mode, a launch control system.

New springs, redesigned electronically-controlled dampers, and bigger front and rear anti-roll bars featured in the LP640.

The basic electronics system of the LP640 mirrored that of the VT in having the same four ECUs detailed previously, but the software was updated and the ECUs modified. These modifications contributed to the LP640's enhanced engine performance and refinement. The new car benefitted from less tyre noise and less gearbox whine.

The permanent four-wheel drive (viscous traction) system – with a 30:70 split front-to-rear drive, but with the ability to instantly transfer 100 per cent drive to either axle and the variable airflow cooling system (VACS), remained largely unaltered in the LP640 variant.

The enhanced vigour from the 6.5-litre engine, combined with the revisions to the drivetrain system, enabled the LP640 to perform the 0-100km/h sprint in 3.4 seconds, and reach a top speed of 211mph (340km/h).

THE INTERIOR

The seats in the LP640 were redesigned for more comfort. They were now wider, with new head restraints. Sport seats were available as an option.

The single stage driver's side airbag contained within the steering wheel centre remained at 60 litre, while the passenger's side airbag was now a two-stage, 130-litre unit.

The leather on the seats, door panels, transmission tunnel, and roof panel featured a new lozenge-shaped stitching called 'Q-citura.' Further leather and carbon fibre customisation was available direct from the factory, through Lamborghini's 'Privilegio' programme. The asymmetric interior, with perforated leather on the driver's side, and smooth leather on the passenger's side, had previously only been available on the Murciélago VT Roadster. This now became an option on the LP640 Coupé.

Although of the same basic layout, the dashboard display graphics were subtly altered.

The LP640 now had – as standard – a Kenwood audio system with a 6.5-inch monitor, that incorporated a radio, DVD player and MP3 player. An optional satellite navigation system could also be ordered at a 2006 price of £1600.

The new LP640 Coupé attracted a price premium of just under ten per cent over the outgoing base Murciélago VT Coupé.

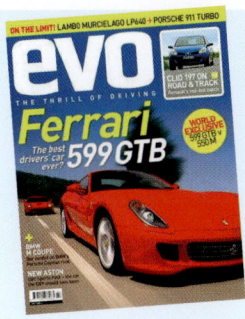

EVO #093, JULY 2006

The Raticosa and Futa passes are a size too small for the Murciélago, narrow and tree-lined, visibility hampered. Even so, the big Lamborghini devours whole sections with no fuss and without even needing to change gear. At low revs the V12 occasionally stutters momentarily, like it's clearing its throat, but by 3500rpm the forces are really building. It means that even tight turns can be negotiated in third, and as they open out you can enjoy the V12's ever-changing voice right up to the far side of 100mph.

That first kick at 3500rpm is followed by a more intense jump in effort at 4500rpm. Now you sense that there really is 631bhp nestled behind your shoulders. Despite the typically long gearing, the revs are piling on here. At around 6500rpm – just when you're fighting the instinct to change up – the engine hits hard again and howls up to 8000rpm. The brick-wall limiter shocks you into an upshift, and with barely a pause to gather your faculties, the ride starts all over again. It's an assault on the senses, your neck pushing against the g-forces, your eyes working hard to process the V12's pure scream ringing in your ears.

The LP640's other trump card is played when you hit the middle pedal and the six-pot callipers bite into the (optional) carbon-ceramic discs. There's no dead play in the brake pedal, so the deceleration hits you hard as the seatbelt cuts into your shoulder, but the LP640 stays composed. Punch some lock into the steering wheel and the front tyres key into the road as the nose slots in towards the apex. Steady the throttle and a bit of understeer does initially build – but it's here that the LP640 departs from lesser Murciélagos.

Stay committed and the understeer stabilises, and as you prepare for the corner exit and start to work the throttle, the car pivots around its gearstick, tightening your line before whipping into a flourish of oversteer as the road straightens. You need to be quick to catch it, flicking on a stab of opposite lock before you can really fill the V12's lungs again. Admittedly it only occurs with the (very intrusive) TCS turned off, but even so, the first time it happens it's a bit of a shock. You could always poke the Murciélago's tail wide, but it took brutality and never felt like its natural stance (at least at road speeds). The LP640, however, is happier to adopt oversteer, but demands you stay on top of it where a standard Murciélago would be covering your mistakes.

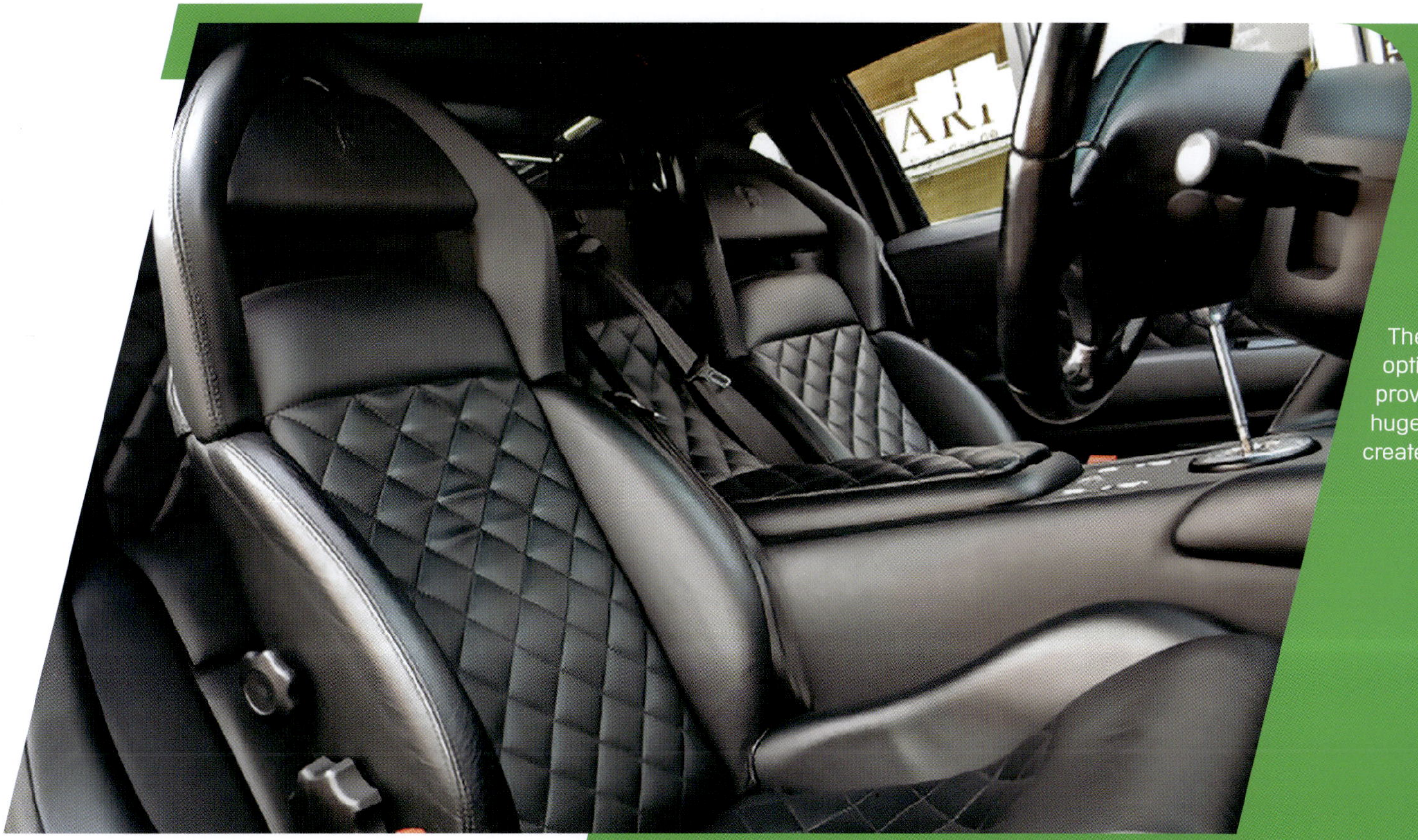

The Q-Citura interior option list is long, providing owners with a huge choice of colours, to create a bespoke interior.

Of course, when that tiny amount of understeer first appears you can back off and keep it neutral, right? Well, you could back off, but then you've got a new problem. Instead of power oversteer, the rear unloads as you come off the gas and the car rolls into momentum oversteer. It's still controllable, but the slip angle tends to be bigger and you need to be careful not to overcorrect. And even with a measured correction, a sideways Murciélago takes up an awful lot of road …

With time you can actively promote the LP640's adjustability, pitching it into clear-sighted corners on a trailing throttle, catching the tail and then unleashing the V12 to overwhelm the rear tyres and ride-out a big, scary and intensely satisfying slide. It sounds more terrifying than it is, as the LP640's four-wheel drive system does its best to assist you by shuffling power to the front wheels as soon as the rears start to spin. Just beware that final injection of power from 6500rpm – when that V12 starts to sing it has a habit of stabbing the tail even wider.

AUTOCAR, 14TH JUNE 2006

The view out, fore and aft, is still scary, the reclined driving position as exotic as ever. Which is why a little respect is due the next day when we bowl out onto the Mugello circuit to drive the LP640 for the first time.

After a steady warm-up lap, I jink through the chicane and accelerate strongly to the final corner. Despite the apocalyptic noise from behind, the LP640 accelerates the way its powerful torque figures suggest: with purpose from 3000rpm, but not quite savagery. Apexing late through the long final corner, I gently squeeze on the power in third and the acceleration starts to build. At 4500rpm the exhaust opens up and the noise begins building to a crescendo. Suddenly we're catapulted forward in a concentrated burst of feral power. Furiously the revs continue to soar as the big Lambo consumes vast chunks of Italian tarmac and humid air. Shifting into fourth only kicks off the whole process again, and the revs falling back right into the sweet spot of the power band. This really is major-league quick, and a proper step up from the old car.

Thankfully, all the launch cars have the optimal ceramic brakes with their 380mm discs and six-pot calipers, so you can brake late and with confidence. Revisions to the gearbox means that in E-Gear paddle-shift form at least (a stick shift is still available), changes now come swiftly and cleanly. There are new springs and dampers, but the basic recipe is the same: a mechanical all-wheel drive system with 28 per cent of drive going to the front axle in normal conditions, but a split of up to 100 per cent either way is possible.

Grip is colossal, the steering a beautiful instrument of feel and precision, and the chassis as communicative as ever. That said, the weight of that V12 behind you still haunts your actions like a ghost sitting on your shoulder. This is a car that demands supreme respect from its driver at all times.

AUTOCAR, 1ST NOVEMBER 2006

To begin with the LP640 comes as quite a shock on the road compared with its predecessor. Not only is it far quieter and more refined than before, but just as Lamborghini claims, there's also far less vibration from the engine than of old. At 70mph on a dual carriageway you'd almost go as far as to say it is civilised.

However, the illusion doesn't last long. Beneath its undeniably more refined personality, the LP640 is still an absolute monster, and this time its bite is bigger than ever. Yet to get the most out of it you do need to rev the engine quite hard, which is amazing considering how many litres of cubic capacity are present.

The key figure to look at is the torque output : 486lb-ft, not developed until 6000rpm. This explains why it doesn't feel quite so rabid as you'd expect when you put your foot down at 2500rpm in either the two top gears. Even so, the 50-70mph time in top gear is still only a nudge above five seconds, which means that the LP640 is actually a fair bit quicker than it feels. Thank the new-found sense of refinement for that, as well as the V12's ability to lug well, from so low down.

Individual, and certainly eye-catching, this Tron-inspired, wrapped Murciélago won't go unseen, night or day!

But what you really want to do is nail it right up to the red line, because only then do you release the full fury of this extraordinary engine's potential. Make sure there's plenty of road to play with before you do, however, because when the LP640 takes off it almost has the ability to bend time. And to get from over here to over there unbelievably quickly.

Unfortunately, we weren't quite able to match Lamborghini's 0-60mph claim; we clocked the car at 3.5sec, a tenth slower than the factory claim and the result, probably, of a slightly less grippy test track surface. Even so, it's fairly obvious from the figures that, regardless of surface, the LP640 is something else against the stopwatch.

To reach 100mph it needed fewer than eight seconds, and to clock 150mph it took 17sec dead; three seconds faster than the 911 Turbo. The standing quarter mile came and went in just 11.8sec with a terminal speed of 125mph.

Amazing as it may seem, this puts the LP640 as close to the McLaren F1 as anything we have ever tested (excluding the Bugatti Veyron), which explains why on the Bruntingthorpe runway it smashed our road test GT top speed record by hitting 194mph. The previous champion was the Ford GT at 186mph.

Unlike its predecessors, the LP640 is not purely about going fast and making as much noise as possible in the process. It's also a remarkably competent tourer, generating far less din than of old from its tyres and gearbox, let alone its engine. Other than the space restrictions of its cabin and near total absence of luggage area, you could happily drive this car to the South of France and back.

The gearbox itself also works far better than before, blipping the throttle quickly and accurately on downshifts and smoothing away upshifts superbly, so long as you help it by lifting slightly during gearchanges. Our only criticism is that it slips the clutch too much when you move away, but then any 1856kg car needs a decent burst of revs to get it moving.

Braking has also taken a big step in the right direction in as much as the LP640, when fitted with the carbon ceramic discs of the test car, stops every bit as well as it goes. Previously, that wasn't exactly the case with the Murciélago.

Driving around the bend: Mercedes-Benz World at Brooklands provides some uninterrupted tarmac for owners during a Lamborghini Club day.

THE MURCIÉLAGO
LP640 ROADSTER

*T*hat there would eventually be a Roadster version of the LP640 could have been taken as a given, by virtue of the sales success and the effusively positive press coverage that the stunning VT Roadster had previously generated.

However, it was only a short nine months following the introduction of the LP640 Coupé that the LP640 Roadster was first shown, on 1st December 2006 at the Los Angeles Motor Show. Here, Lamborghini gave due prominence to California as one of the most – if not *the* most – important markets for this open-top barchetta, with a difficult-to-fit manual canvas emergency roof. Six days later, the Roadster made its European debut in rather more rainy northern Italy, at the Bologna Motor Show.

This new Roadster was essentially an amalgam of the Roadster VT and the LP640 Coupé, and, to avoid repetition, the shared aesthetic and mechanical details of this car can be found in those two chapters.

To summarise, the LP640 Roadster inherited the engine, drivetrain, transmission, chassis, suspension and electronics of the LP640 Coupé, as well as its new front and rear bumpers, asymmetric sill air intakes, new rear diffuser with integrated single exhaust outlet, and new tail lights. These were then married to the key Roadster VT features of a more steeply raked windscreen, lower ground clearance, the loss of the steel roof and its replacement with fresh air and unlimited headroom or a canvas tent, and the spectacular rear-hinged engine lid. The asymmetric use of perforated leather on the driver's side and smooth leather on the passenger's side, which was a key interior design feature of the VT Roadster, continued as standard in the LP640 Roadster. More extreme customer taste requirements could be met through the 'Ad Personam' customisation service offered by the factory. Less obvious, but rather more importantly, the new Roadster retained its predecessor's electronically-controlled and activated hidden roll-over bars, which could shoot up from behind the seats in a few milliseconds, should an imminent roll-over situation be detected.

Contemporary motoring reports almost unanimously celebrated the beauty of this new Roadster, and it did indeed have an even more arrow-like profile than the VT version, mainly due to its more prominent front spoiler. With softer suspension settings than the Coupé, this Roadster – like its predecessor – was said to provide a better ride than its Coupé sibling, and was better able to cope with road surface imperfections. Again, as with the VT version, there was much less chassis flex than would be expected in a car in which a key structural element had been removed Additional cross-bracing, including new carbon fibre members, largely compensated for the loss of the steel roof. The expected, and legitimate, criticism of the soft-top was again expressed, and again too harshly and in too many words, bearing in mind that the various Murciélago Roadsters were always designed as classic barchettas.

In the United Kingdom, the base LP640 Roadster debuted with a price of about £200,000, when the list price for the departing base VT Roadster and the base LP640 Coupé were both about £190,000.

EVO #101, FEBRUARY 2007

We take the Roadster to the same roads where we tested the Coupé LP640 six months ago (093). The Raticosa and Futa passes south of Bologna are narrow, winding and riddled with tightening turns and testing, bumpy apexes to catch out the unwary. As a supercar test route it's as tough as they come; if there are any weaknesses with the Roadster's chassis, we'll discover them here.

The steering requires a firm hand, particularly through hairpins, but, as before, it lightens at speed, feeding a high level of confidence-inspiring detail back through the thick suede wheel and allowing accurate placement – vital in such a wide machine.

The springs and dampers have been revised from those fitted to the Coupé LP640, however the resulting handling characteristics remain very like those of the Coupé we drove on eCoty (evo 099), suggesting there was something wrong with the first Coupé which we drove here. That car was a nightmare – very edgy in corners, permanently on the tips of its toes. The Roadster feels much more benign, much more settled and driveable. Its set-up allows for a mid-turn rethink should it prove necessary. Need to tighten your line? No problem. Push hard and the nose eventually begins to scrub wide – and if you leave the traction control switched on, this is as far as things will develop, the

electronics neatly keeping everything on the curved and narrow.

Switch the same system out and initially the results are the same, with the gentle understeer building further and the nose running wider still. Back off now and the Pirello P Zeros recover their hold on the tarmac without fuss. Stay on the power, however, and things rapidly become very exciting indeed. The 13in-wide rear tyres are finally overwhelmed by the onslaught of power and flick wide, and when it happens you'd better be ready to deal out some rapid and decisive corrective action, for there's a whole lot of mass on the move. Don't panic and jump off the power, though – the sudden weight transfer will accentuate the slide to dramatic, road-filling proportions. Instead, balance the throttle and the LP640's four-wheel drive system will divert more motive energy forwards, giving the front added bite and pulling the Roadster clear from the slide.

Drive just under this rather hairy zone and the alfresco Lambo flows over these near-deserted mountain roads, each pull of the right-hand (optional) e-gear paddle unleashing another savage burst of inexorable thrust. The (again optional) carbon-ceramic brakes with six-pot callipers perform tirelessly all day, repeatedly hauling off speed with a firm and consistent feel, allowing you to develop a rhythm similar to that in the Coupé and cover the group at a similarly scorching pace.

Out here, when you're wrapped up in the drama of it all, the small degradation in structural rigidity and the tiny weight penalty count for nothing. Coupé or Roadster? It all boils down to whether you like driving occasionally in the rain.

AUTOCAR, 20TH/27TH DECEMBER 2006

This is the Roadster version of the Murciélago LP640 launched six months ago, and given that it brings 631bhp and 486lb-ft to twist Dali-esque shapes through the 1665kg chassis, you really scratch your head as to why it's so rigid. It has no right to be.

The trick is that the Murciélago's chassis isn't a monocoque with a hole in it. It's a spaceframe which already has lots of holes in it. To build a Roadster, they just made the hole around the cabin bigger and put in 95kg of bracing – including a massive frame around the engine – to compensate.

It seems to work, and it needs to, because Lamborghini gives the Roadster no let-up from the 6.2-litre V12's brutality. It still fires up with an air-shredding *braaap* before settling to its lumpy idle.

Pluck first in the optional E-Gear and it trickles off without histrionics into the Sant'Agata morning. Its immediately louder than the Coupé, even at idle, but it's the part-throttle action that makes you warm to it.

Bulls of a feather, flock together during an event at Silverstone in 2010. There's a great network of owners and enthusiasts within the Lamborghini world, so support and like-mindedness are never far away.

It's got a catalogue of gurgling noises on downshifts and it pops and crackles.

While the towns and cities are its life's blood, its not exactly practical; you cannot see a thing past 90 degrees to the car. The next thing you notice at these gentle speeds is the ride quality. Find a lumpy road to check for scuttle shake and it'll surprise you. Yes, the wheels drop into holes, but it is not the spine-thumping, endless array of vertical inputs, which is what happens when you're in the Coupé. Instead, you aim for big holes and feel, well, not much. It's very, very together.

That's not to say there's no scuttle shake. There is. Lots of it. The engine cover either oscillates sideways by about two inches or the mirror and screen move a lot. Maybe it's both.

Yet, somehow, this shaking has little impact on what is happening below decks. There's an occasional jitter through the steering column, but everything else just seems like a softer, friendlier Coupé.

But use it in anger and it will take your breath away. The chirping snap of the four throttle bodies is the first assault, then the howl of protesting air molecules fighting for space as they're rammed into tortured inlet tracts. The E-Gear is pretty good too, and it snaps

Romanian LP640 Roadster at the Col du Turini, with my series 1 MX-5 in the background.

At Geneva in 2009 – and sporting a rather nice aftermarket hard-top. Shades of Reventón, perhaps.

the shifts through so quickly you barely notice the lack of delicious metallic snicking that comes from the manual gate.

But the chassis is the surprise. It isn't the soggy listless bucket you would expect. It's not nasty and it's not dull. There is almost as much steering feedback as in the Coupé, and its reasonably accurate. It will even change direction with conviction and with a way of masking the high centre of gravity that the Coupé hasn't yet mastered.

Styled by Bob Forstner, and dripping in carbon fibre from the front splitter to the roof, this LP640 makes a bold statement. The Hermera wheels compliment the overall style.

THE MURCIÉLAGO
LP670-4 SUPERVELOCE

To the purist, the Lamborghini Murciélago LP670-4 SV is not only the pinnacle of the Murciélago range, but is also of some historic importance, in that it houses the final and ultimate iteration of the classic Bizzarrini V12 engine.

At the press conference following its unveiling on 3rd March 2009 at the Geneva Auto Salon, the then-Chief Executive of Automobili Lamborghini SpA, Stephan Winkelmann, said of the two-seater berlinetta: "The new Murciélago LP670-4 SuperVeloce is the systematic continuation of our brand philosophy. It is more extreme and uncompromising than virtually any other automobile. As the new top model of the highly successful Murciélago range, the SuperVeloce displays not only outstanding driving dynamics, it is also further evidence of our company's technological expertise. Customers can look forward to an utterly unparalleled driving experience."

At its launch, Lamborghini declared that the LP670-4 SV would be a limited edition car, with a production maximum of 350 units. In the end, only 186 units were produced.

While the reasoning behind the first part of this new car's name can easily be guessed from that of its predecessor – 'LP,' or 'Longitudinale Posteriore,' refers to the engine positioning within the chassis, the '670' refers to the uprated engine's maximum output in horsepower (equating to 661bhp), and the '4' refers to the four-wheel drive Visco-Traction system – it is the 'SV' appellation that will most excite the purist.

'SV' stands for SuperVeloce, which translates to super or extreme speed. The SV label has been used very sparingly by Lamborghini, and then only on its most special models. At the time of the 670-4 SV's release it had appeared on the Diablo SV, but more notably had originally been used on the most desirable variant of what many consider to be the first supercar.

The Miura SV was the ultimate (if you disregard the very rare Jota cars) expression of the ultimate super sports car of its time. Between 1971 and 1973, only 150 Miura SVs were produced. Of course, the 60-degree Bizzarrini V12 featured here, too, but in this instance it was positioned transversely, and was claimed to put out 380hp at 7700rpm, and 286lb-ft at 5500rpm, allowing the 1298kg SV to accelerate from rest to 60mph in about 6.5 seconds, and to reach a top speed of about 175mph. Today, the Miura SV is one of the most coveted, most collectable, and most expensive of all cars.

The Diablo SV was also introduced at the Geneva Show. At its debut, in 1995, it was the cheapest of the Diablo range, but also the purest; it combined two-wheel drive with an increased power output of 510hp. Its most notable aesthetic feature was the standard equipment adjustable rear spoiler, while its most notable mechanical feature – other than the modified engine – was its larger diameter (340mm) front brake discs.

WEIGHT SAVING

The swan-song of the Murciélago range and the Bizzarrini V12, the LP670-4 SV aimed to be special through a combination of meticulous weight saving and the addition of more power and torque.

One obvious way of dramatically cutting weight would have been to abandon the four-wheel drive system in favour of a simpler, cheaper, and lighter two-wheel drive system. This approach was rejected for a number of reasons. Firstly, it was felt that the increased power of the 670-4 SV could not be effectively transmitted to the tarmac through a two-wheel drive system. Secondly, the added traction provided by the front wheels in a four-wheel drive system was considered to be an advantage in accelerating out of tight bends. Thirdly, the four-wheel drive system provided an extra layer of safety when driving this high torque output car. There might also have been an unspoken fourth reason, in that Audi, Lamborghini's parent company, was heavily committed to the Quattro four-wheel drive system, and abandoning such a system in favour of a two-wheel drive system for the ultimate expression of the ultimate super sports car would have been bad marketing.

Instead, an abundance of carbon fibre, the jettisoning of unnecessary equipment, and the use of lighter and stronger materials wherever possible were seen as the keys to losing kerbweight.

To this end, the classic spaceframe chassis in the 670-4 SV now used ultra high strength sectional steel, which meant that thinner tubing could be used, saving a total of 20kg from the spaceframe alone. An additional advantage that accompanied the use of this higher strength steel was that the torsional stiffness of the spaceframe increased by 12 per cent, which benefited handling.

Alcantara microfibre is lighter than leather, so this was adopted for most of the interior trim. Lighter forged alloy wheels were sourced. Magnesium is the lightest metal found on Earth, and is also the second most abundant metal in the Earth's crust. It is 33 per cent lighter than aluminium, and 75 per cent lighter than steel, but is stronger per unit volume than either. All possible aluminium structural components and brackets in the LP670-4 SV were therefore replaced with magnesium ones.

The Kenwood multimedia entertainment system, standard issue in the LP640, was ditched in the name of weight-saving, although it could be added on as a no-cost option.

The motor for the electronically-controlled rear movable spoiler was removed, saving further weight, and a choice of one of two fixed rear carbon fibre wings was offered to the customer instead.

Unusually, with this car Lamborghini specced the E-Gear paddle shift transmission as standard, because this system weighed less than the six-speed manual gearbox system. A manual gearbox could, however, be ordered as an option. Apparently only about six customers chose the manual transmission option. A new lightweight clutch was used in both variants.

But it was in the lavish use of carbon fibre that Lamborghini really went all out, when designing and constructing this very special Murciélago. Carbon fibre was used in, amongst other things, the front and rear bumpers, the lower sill air intake covers, the rear wing and supports, the casing for the third brake light, the seats shells, the floor, the transmission tunnel and the door skins.

These modifications resulted in an overall weight reduction of 100kg compared to the LP640 Coupé – which breaks down to 33kg removed from the chassis and exterior, 34kg removed from the interior and 33kg removed from the engine and drive train.

EXTERIOR MODIFICATIONS AND AERODYNAMIC ADDITIONS

Optimisation of the LP670-4 SV's aerodynamics dictated several striking bodywork modifications, which immediately distinguished it from its lesser brethren, particularly when the matt black painted carbon fibre additions contrasted strongly against a vivid body colour.

The front end of the car now had an even more prominent spoiler,

The engine cover is a piece of art in itself, and follows a traditional Lamborghini style that was first seen on the likes of the ground-breaking Marzal from 1967 (inset) and the hexagonal shapes have been a defining design feature on many of the cars ever since .

attached to the front bumper by two vertical carbon fibre struts, so giving the appearance of three front air intakes. The sharp angle of the bumper, spoiler and the struts, as well as the continuation of the spoiler laterally across the side of the bumper, all contributed to a particularly aggressive front and side view.

The principal feature of the side view was now the black painted carbon fibre covers for the lower sill air intakes, which remained asymmetrical from the right side to the left.

The back view was dominated by a fixed carbon fibre rear wing, which was supported by two prominent carbon fibre struts. Two different rear wings were available; the standard item was a relatively small wing, but a massive wing which Lamborghini called the Aeropack Wing was also available as a $7000 option. The Aeropack Wing added substantial downforce at high speed, but carried with it the penalty of added drag, reducing the car's top speed by 3mph compared to the standard smaller wing.

The back of the LP670-4 SV also featured a large full-length double-decker rear diffuser which was bisected by the two carbon fibre struts supporting the rear wing. The upper of the two diffuser bars was indented by the huge, wide and flattened single exhaust pipe. Immediately above the exhaust pipe sat a large, laser-cut and Teflon-coated hexagonal aluminium mesh to allow hot engine bay air to escape. Above this, and between the distinctive tail lights was a black panel unnecessarily carrying the italicised Lamborghini logo – just what other marque could present such a rear?

The engine lid was also substantially modified, with a black carbon fibre framework supporting a transparent centre through which the engine could be seen. This transparent centre was actually made up of three descending terraced plateaus, like a traditional Balinese padi field. Each terrace was made up of a transparent hexagonal polycarbonate polymer plate, which was open at its rear end to help engine ventilation.

The LP670-4 SV could be ordered in one of five standard colours: Giallo Orion, Arancio Atlas, Bianco Isis, Grigo Telesto and Nero Aldebaran, or one of two extra-cost optional colours: Bianco Canopus and Nero Nemesis.

THE INTERIOR

The interior of the LP670-4 SV is largely made up of Alcantara and carbon fibre. Sport bucket seats with a carbon fibre shell and an Alcantara covering come as standard, and feature a Y-shaped, stitched, body-coloured design running through the centre. The inner door panels and the transmission tunnel are made of carbon fibre, and Alcantara dominates the interior. Both these materials contribute a sporty look and feel to the cockpit. The audio-navigation system

– which was deliberately left out of the LP670-4 SV 's standard specification – could be reinstated as an option.

THE MECHANICS

The final and ultimate Bizzarrini-derived Lamborghini V12 engine in the LP670-4 SV has aluminium alloy construction, the ideal 60-degree cylinder angle, dry sump lubrication, and four chain-driven camshafts. A maximum of 670hp (661bhp) and 660Nm (487lb-ft), produced at 8000rpm and 6500rpm respectively, was coaxed out of this 6496cc unit, thanks to changes to the intake system, valve train optimization and – most importantly – increased travel for the variably timed valves. The intake system was now of variable geometry, with three operating modes. A completely new exhaust system was designed for this car, with a new back-box, and the large central tailpipe was made of a new and lighter material than in previous variants of the Murciélago.

The standard issue E-Gear transmission in the SV has both a Corsa (Race) mode and a Low adherence mode available. The classic open gate six-speed manual transmission was available as a no-cost option, and both featured a new lighter clutch with a disc diameter of 272mm.

The transmission ratios were as follows:

The LP670-4 SV has a front-to-rear weight distribution of 42 per cent to 58 per cent, and its four-wheel drive system sends up to

First gear	3.091:1
Second gear	2.105:1
Third gear	1.565:1
Fourth gear	1.241:1
Fifth gear	1.065:1
Sixth gear	0.939:1
Reverse gear	2.692:1
Final drive ratio	2.53:1

35 per cent of the engine's torque to the front wheels, until the rear wheels start to slip, at which point up to 100 per cent of the available

torque can be sent to the front axle. The front and rear limited-slip differentials were set at 25 per cent and 45 per cent respectively.

The rest of the mechanics follow the classic Murciélago recipe, with double wishbone suspension at all four corners, and front axle lift was standard on the SV. Carbon Ceramic drilled and ventilated brake discs measuring 380mm all-round, with six-piston callipers, were also standard, and provide excellent and fade-free retardation to this 1565kg road missile. The Variable Air-flow Cooling System, the car's electronics systems, and the airbag systems remain largely unchanged.

Lamborghini claimed the following performance statistics for the LP670-4 SV:

Top speed with the standard rear wing	212mph
Top speed with the larger aeropack rear wing	209mph
0-100kph	3.0s
0-200kph	10.80s
0-300kph	30.80s

The list price of the base LP670-4 SV in the United Kingdom in the summer of 2009 was £221,335, which was about 18 per cent more than a LP640 Coupé – a well justified premium for the technological improvements, the exclusivity, and the historic nature of this ultimate Murciélago variant.

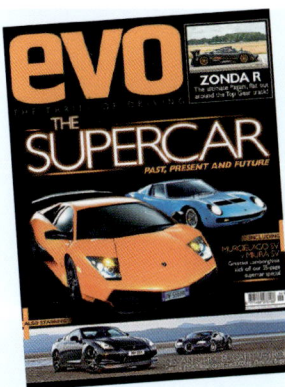

EVO #134, SEPTEMBER 2009

Don't think that driving the Murciélago is as simple as driving a Mondeo because it's not. You feel your concentration spike every time you drop down into the seat (which actually needs a bit more lateral support) and turn the key. The forces that it exerts on you under acceleration, or when you hit the huge carbon brakes (they still need more progression, but they're much better than in the Gallardo) or when you lean on the Pirelli P Zero Corsas are massive and intimidating. To keep your foot pinned on a valley road, accelerating unremittingly as you flick each gear at the limiter and watch the strip of tarmac in the windscreen narrow (continued P114)

That aggressive rear end – the view that most other road users see.

Profile is 100% Lamborghini, and there's no disguise when it comes to advertising the 'Superveloce' ranking.

Even standing still, the LP670-4 SV looks as if it's flying.

Driver's eye view - Beautifully and practically laid-out, the SV's instrumentation tells drivers everything they need to know, when it is required. A speedometer that goes to 360 km/h, for example, is essential! Right-hand drive versions show 220mph. Of course, it may well be time for a new mortgage application and a service if this many lights stay on ...

Alcantara looks fabulous; especially when juxtaposed against carbon fibre with a high gloss finish. The contrast stiching on the sculpted seats, unique to the SV, ensures continuity of the SV branding – an interior of extremes, echoing the soul of the car.

relentlessly around the car as the speed builds, is to be immersed in an unforgettable supercar experience. But I like the fact that it still feels planted at 170mph, not flighty and lethal. It makes me want to go back to the 170mph and experience it again, push further.

And in the corners it is even more noticeable. On entry you can brake so late you're weightless in the seat belt. Turn in and you can adjust the balance through the corner. Because it corners flatter and there's more edge to the grip, the nose of the LP670 actually pushes more obviously than the LP640. As a result you need to play sensitively with the throttle and steering through a long corner, leaning into the invisible lateral g but feeling the fluctuations in load as you adjust your inputs. Bumps are soaked up and telegraphed by the SV's suspension in a way you can compensate for. You don't jump on the throttle early in the corner because the big rear tyres will push the nose wider still, so you wait until you're past the apex, then you begin straightening the wheel and feeding in the power, using the traction of the rear-biased four-wheel drive to ride the furiously ignited torque. You can't turn the ESP to Corsa mode and take ham-fisted liberties – there's still a huge V12 slung behind you. But even on the road you can dig right into the Murci's handling repertoire and enjoy it – not just nibble nervously at the edges.

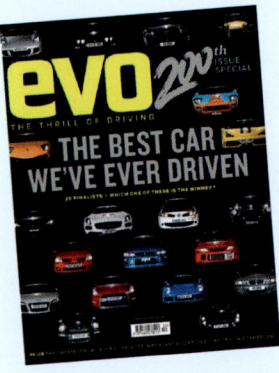

EVO #200, OCTOBER 2014

The Lambo buzzes with visual energy, draws you closer in with fascinating sculpted details and then delivers an uppercut with its scissor door. No one does wild quite like Lamborghini and the LP670 was the fitting final send-off for the Murciélago: lighter, more powerful and more exciting than any before.

Then there's the engine, always the engine. Real supercars have 12 cylinders and a soundtrack that's a thrilling sonic landscape. Select third gear, find a long-ish straight and, loping along at tickover, floor the throttle. The Lambo's 6.5-litre V12 snaps you forward instantly, initially churning out a low, heavyweight pulse that ripples through your soft tissues. It then evolves, becoming gradually lighter and more urgent until at somewhere between 5000 and 6000rpm the fizzing fuse hits the kegs of gunpowder. *Pow!* The engine note loops upwards and the SV lunges forward, covering the last couple of thousand rpm so quickly that the first few times you're too late to prevent it battering the limiter. This particular SV, generously loaned by affable owner Andy Peirson, is extra special as it has been factory-tweaked up to 700 horsepower. It's a claim I wouldn't dispute, having felt out it soars, clean and pure, to the 8200rpm red line.

There's a rather old-school feel to the rest of the SV, particularly the transmission. Its automated single-plate clutch can be rather clunky if left to its own devices and is simply brutal in Corsa mode. Meanwhile, accessing the considerable reserves of the Michelin Cup-shod chassis means gritting your teeth, so beefy is the steering's weight. The SV delivers unmatched visual, aural and performance thrills, but it's not the most compelling driver's car here.

AUTOCAR, 19TH JULY 2009

The SV's personality is akin to that of the bloke who turns up a party with two supermodels on each arm, and leaves with another three in tow. Yet beneath its He-Man exterior, bolstered in the case of the test car by a comically enormous optional carbon fibre rear wing, the SV also showcases how hard Lamborghini is thinking technically nowadays.

Despite its seemingly vast dimensions and aspirations, this car weighs little more than 1600kg, thanks to the extensive use of carbon fibre and other lightweight materials.

So given that the 6.5-litre engine produces a thundering 663bhp at 8000rpm and 487lb-ft at 6500rpm, it's hardly surprising to discover that the performance is quite a long way the other side of impressive. Think 0-60mph in 3.2sec and 0-100mph in under seven and only then will you get an inkling of what it feels like to open the accelerator in second gear and hold it there for a while.

On the road the SV feels suitably terrifying and has a pretty uncompromising ride, even for a Lambo. But it's also one of the most exciting experiences you'll ever have on four wheels.

The steering is notably sharper than in a regular Murciélago, and the extra agility during direction changes makes it feel both lighter and less clumsy when you're really going for it.

You can sense the reduction in weight more than the increase in grunt in virtually everything the SV does, right down to increased power and response under brakes. Dynamically it takes the game at least one notch forward compared with the LP640. Which is no mean achievement.

Should I buy one? If you are the sort of person who a) has the financial clout to do so and b) prefers the idea of cars like the Ferrari F40 as opposed to the Bugatti Veyron, then yes.

The new Murciélago SV may not quite be the fastest or most civilised supercar on the road, but it's one of the most exciting there has ever been. It's a fantastically gutsy statement from Lamborghini.

THE REVENTÓN

*W*as the Reventón Lamborghini's Centro Stile's first attempt at establishing itself as a haute couture atelier?

First used in reference to the bespoke clothing made by Englishman Charles Frederick Worth, at his Paris fashion house in the period between 1850 and 1880, haute couture's literal translation into English reads as 'high dressmaking.' The use of the term haute couture is now protected by law in France, and a fashion house can only apply this term to itself if it meets the following three criteria: firstly, it must design bespoke clothing for private individuals, which involves at least one personalised fitting session; secondly, it must have a workshop in Paris employing at least 15 full-time staff; and finally, it must present a collection of at least 50 original designs to the public twice a year, in January and July.

The truth of the matter is that the Lamborghini Reventón is a gloriously attired Murciélago LP640, but this should not detract from the fact that it had purpose. First, it showcased Lamborghini's in-house styling talent. Second, it was a design pathfinder for future cars. Third, it acted as a medium through which Lamborghini's Centro Stile could closely collaborate with Lamborghini's Research and Development Department. Fourth, it was a development test bed for bringing new technology from the concept and design phase through to production as quickly as possible. This was the first production Lamborghini to go from sketches to CAD (computer aided design) to production – a process that Lamborghini hopes to adopt for its future cars. In the Reventón's case, this took less than 12 months. Fifth, with a 2007 price-tag of one million Euros, before local taxes, and when a base LP640 was priced at one-fifth that amount, it was intended to generate actual profit for Lamborghini. Finally, by strictly limiting the production of the Reventón Coupé to just 20 units, Lamborghini sought to emulate one of Ferrari's most successful business plans – produce a highly exclusive product, and only offer it to your most favoured, most loyal, most celebrated customers, and thereby elevate the desirability of the whole marque. In line with this plan, every one of the 20 Reventón Coupés was sold prior to production.

In the finest Lamborghini tradition, the Reventón was named after a famous fighting bull, which carried – to its credit or discredit, depending on your views on bull-fighting and animal rights – the honour or dishonour of having gored the famed matador Felix Guzman to death in 1943.

The Reventón, at the time of its debut, was supposedly the most expensive production (though, admittedly limited volume) car ever put out by the famed Sant'Agata manufacturer. Manfred Fitzgerald was the Head of Design at the time of the Reventón's gestation, and he very clearly stated that this was an entirely design-led project.

The mechanicals of the Reventón were pretty much unchanged from the LP640. The engine was merely a blueprinted version of that present in the LP640, and through careful polishing, porting and balancing of the engines internals, an extra 10bhp was liberated, bringing the total output up from 631bhp to 641bhp; a pretty inconsequential power increase in a 1695kg car. The same six-speed paddle shift gearbox, tubular steel chassis, braking system, active cooling system and electronic control systems were all carried over from the LP640. The 6496cc normally-aspirated dry sumped V12 in the Reventón produces a maximum of 650bhp at 8000rpm, and 660Nm at 6000rpm, which, together with the standard E-Gear transmission system and the viscous traction permanent all-wheel drive system, allows this car to accelerate from rest to 62mph in 3.40 seconds, and reach a top speed of 211mph.

No, the Reventón stands out for its interior and exterior aesthetics. A survey conducted by Lamborghini of its most valued customers had shown an unmet demand for a highly exclusive car that was also reliable. Lamborghini therefore made no excuses for basing the Reventón on the tried-and-tested LP640 platform.

For inspiration, the entire Lamborghini Centro Stile staff spent a day at an American air force base in northern Italy, examining fighter planes. It was the F-22 Raptor fighter that finally stole their collective imagination, and served as the design basis for the Reventón.

The Reventón is highly angular, with extremely sharp-edged creases, and is even more arrow-shaped than the LP640. The triangle stands out as its design motif. This car retains the LP640's spaceframe chassis with carbon composite reinforcement, and its new angular carbon fibre body is glued and riveted to this traditional steel frame. It does not share a single body panel with the LP640, and the only immediately identifiable exterior part in common is the wing mirror.

At the front, the Reventón adopts an even more distinctive nose-forward appearance, thanks to the redesigned, sharply protruding front bumper. There are now three massive, sharply creased front air vents, and the sharp-tipped front spoiler is angled even more acutely backwards towards the lower bodywork, further accentuating the arrow theme of the Reventón.

The front wings are also radically changed, to accommodate the new headlight covers. These are now five-sided, and appear to be inverted and narrower versions of the covers already present in the Murciélago – inverted in the sense that the widest section of the cover is now in the lowest and most forward position, with the highest and most rearward section narrowing almost to a point. The front and rear lighting systems have also been modified to incorporate light-emitting diodes (LEDs), which now sit alongside the traditional Bi-Xenon bulbs in the headlights. There are seven daylight-running LEDs, and nine indicator and hazard LEDs at the front, and the rear LEDs are new special heatproof items, able to withstand the heat from the nearby V12 engine. All of the rear driving, brake, indicator and hazard lights feature these new heatproof LEDs.

At the sides, the front side indicators have been changed to a broad-based pyramid, and the lower air intakes along the side sills (continued P120)

Main picture: Alcantara, carbon fibre and extruded aluminium in perfect harmony – the interior of the Reventón.
Right: Stealth-like in its styling. The design team spent many hours studying and researching Aeronautical design; the influence and attention to details, such as the carbon fibre elements of the wheels and the fuel filler cap, are clear.

From every angle, the Reventón pleases the eye, and, with hindsight, the influence on the Aventador is plain to see — but which one? Roadster or Coupé?

are enlarged, more angular, and feature additional strakes as well as a double outlet vent cover. The vents around the active variable airflow cooling system have also been modified.

The rear bodywork presents a much more aggressive appearance due to its sharp angles. In the same way that the front bodywork appears to thrust forwards, the rear bodywork appears to thrust backwards. This mirroring effect is further emphasized by the pointed lower bumper and the huge angular rear vents. There is a prominent rear diffuser, above which now sits a single, slightly curved exhaust pipe of considerable size. The tail lights are narrower and set within angular carbon fibre frames. The engine lid, made of carbon fibre composite with glass laminate panes, allows a clear view of the engine. The glass panes are arrow-shaped in keeping with the Reventón's design theme, and open at the rear to allow better engine heat venting.

The Reventón features a dramatic new wheel style with a five spoke 'Y' design. Opaque carbon fibre fins are actually screwed onto the black aluminium spokes. These fins create a wind turbine effect, drawing in air to further cool the ceramic brake discs hidden behind the wheels. The standard carbon ceramic brakes, with each disc measuring 380mm in diameter, together with six-piston callipers, ensure massive and fade-free deceleration. The Reventón's 18-inch wheels carry 245/35 tyres at the front, and 335/30 tyres at the rear.

Lamborghini developed a new paint for the Reventón, also called Reventón. This is military-style grey-green, essentially matt in finish, but given additional depth and glitter due to the metallic particles embedded within the paint. Every one of the 20 Reventón Coupés left the factory in this exclusive colour. Another example of the care taken with the aesthetics of this car can be seen in the fuel tank filler cap. It is milled from a single block of aluminium, and is as much a delight to look upon, as it is to touch.

The interior, and particularly the dashboard display, are stand-out features of the exclusive and expensive Reventón. The interior is not dissimilar to the standard LP640 in basic design and layout, but has an additional abundance of carbon fibre, Alcantara and aluminium. The unique Alcantara-covered sport seats are more bucket-like, and the carbon fibre dash binnacle is also exclusive to the Reventón. The steering wheel is Alcantara covered, except for the bottom, which is bare carbon fibre.

Other than the striking new bodywork, which was designed by Alessandro Serra, the other stand-out feature of this car is the stunning dash display, that takes its inspiration from military fighter jet planes. The display instrumentation is made up of three Digitech Thin Film Treatment (TFT) liquid-crystal displays, housed within an angular, aluminium milled binnacle, which is itself covered by a carbon fibre outer casing. Lamborghini holds patents on this novel TFT display system. The driver is able to select between two different display modes at the touch of a button: either a quasi-analogue display with circular dials, or a much more radical display utilising two converging lines simulating a pilot's eye-view of a landing strip, to mark out engine and road speeds using illuminated bars. Another innovation is the central g-force meter, which displays both longitudinal and lateral g-force, as a function of acceleration, braking and cornering.

THE REVENTÓN ROADSTER

The Reventón Roadster made its debut at the 2007 IAA in Frankfurt, where the President and CEO of Lamborghini SpA, Stephan Winkelmann, introduced it with the following words: "The Reventón is the most extreme car in the history of the brand. The new Roadster adds an extra emotional component to our combined technological expertise – it unites superior performance with the sensual fascination of open top driving."

The Reventón Roadster is very much an amalgam of the Reventón Coupé and the Murciélago Roadster, and shares many features derived from both these models. The following text therefore only highlights those features unique to the Reventón Roadster, without repeating in detail those common aspects that have already been described in the Reventón Coupé and the Murciélago Roadster sections of this book.

Each Reventón Roadster was priced at a ten per cent premium over the closed top version, with a list price of 1.1 million euros, before local taxes. Only 15 of these cars were built, making the Roadster 25 per cent more rare and exclusive than the already highly-exclusive Coupé version. This car inherits the Lamborghini Murciélago LP670-4 SV's engine, which produces 661bhp at 8000rpm and 487lb-ft at 6000rpm, to launch the Roadster to 62mph from rest in 3.40 seconds, but the top speed is reduced by 6mph, to 205mph, compared to the Coupé version. The Roadster has a dry weight of 1690kg, which is 25kg more than the Reventón Coupé. It retains the Murciélago Roadster's rudimentary canvas roof, its active roll-over bars, which can pop up from behind the seats in milliseconds, and its lattice-like engine cage to compensate for the loss of chassis rigidity through the lack of a solid roof. The Roadster features another unique matt grey paint finish, called Reventón Grey, which again has metallic particles impregnated within it to give further depth. There is now a fourth brake light incorporated into a new and prominent cross member that spans the top of the engine lid. The Alcantara- and carbon fibre-clad cockpit retains the revolutionary TFT LCD displays first seen in the Coupé.

All these design and mechanical modifications culminate in a barchetta whose only genuine period rivals were the open top versions of the Bugatti Veyron and the Pagani Zonda F. Both these cars have significantly better power-to-weight ratios, but the Reventón Roadster is still a formidable opponent from a purely aesthetic viewpoint.

Unmistakably Lamborghini: the futuristic yet almost timeless design of the Reventón.

AUTOCAR, 14TH NOVEMBER 2007

But even while you are playing with the Reventón's new dash, you are being provoked by a catalogue of deep burbling noises from that 6.5-litre engine. Move off slowly and you'll be saturated by the pops and crackles coming out of the unit on downshifts (or even on lazy upshifts), and the pops and crackles of the continuously variable inlet and exhaust timing as the ECU keeps it from drowning.

A paddle-shift gearbox that hates and jerks its way through the multi-point turns demanded by its wide turning circle is probably the lowlight.

Ride quality is not the Reventón's thing either, mind you. It's better at speed, but around town the combination of taut thinly padded seats, wafer-thin sidewalls and suspension built for violence means that cobblestone streets are best avoided. Things get better with pace, though. The same bump that braced you at 10mph is barely felt at 40mph. The wheel still drops into the hole, but it is no longer a spine-thumping, endless array of vertical inputs. The faster you drive it ,the more together the Reventón feels.

Our car, an early demonstrator, was limited to 80mph, which is just rude. You can slam past that in second gear. There's brutality to everything it does, but it's also extremely flexible, with a lazy ability to twist driveshafts from the tiniest throttle openings.

You sometimes wonder if that's its preference, but when you snap the throttle open, the motoring world has few better, more tingling sounds of anticipation than its four throttle bodies chirping and hissing open, than the roar of tortured induction air over your shoulder.

It's stable mid-corner as well, provided you're not being ham-fisted with it.

The Reventón's architecture is so old that electronic stability control couldn't effectively be retro-fitted, so it relies on all-wheel drive and traction control to stay on the road. If you're not completely concentrating, that's not enough, because in the wet it can get wheel-spin from all four tyres, even in fifth.

There are no doubts that it is indeed capable of the claimed 3.4sec 0-62mph sprint, and the Reventón's slipperier shape makes the 211mph top speed seem a tiny bit pessimistic.

But it's the impact, rather than the driving experience that defines the Reventón. If you were just buying the driving experience, after all, you'd take five LP640s for the same money.

MURCIÉLAGO 40TH ANNIVERSARY

In 2003, as a 40th birthday present to itself and its customers, Lamborghini produced just 50 Murciélago 40th Anniversary units, each of which was accompanied by a personalised certificate. Painted in an exclusive three layer pearlescent jade green colour called Verde Artemis, the car ran on contrasting grey anthracite-coloured wheels, within which hid silver-grey brake callipers. In the 40th Anniversary, the body part directly in front of the variable airflow cooling system

vent was made of carbon fibre, with further carbon fibre extending forwards to the side glass support. The car had a 40th Anniversary logo on its side, and had a modified exhaust system, but was otherwise mechanically standard.

On the inside, the driver's side featured grey perforated leather called Grigio Syrius, while the passenger's side was cloaked in smooth black leather. The floor mats carried the 40th Anniversary logo, and each car also came with a fitted carbon fibre suitcase.

'Nineteen of fifty,' – a very appropriate plate, in Roman numerals, accompanies this Canadian 40th anniversary model.

MURCIÉLAGO 640 VERSACE

The Lamborghini factory 'Ad Personam' programme was overshadowed by a partnership between the Gianni Versace fashion house and Lamborghini SpA, with the debut of the Murciélago 640 Versace edition at the 2006 Paris Motor Show. The production run was limited to only 20 units, and there were only two colour options – a white 'Isis' edition, and a black 'Nero Aldebaran' edition. Ten units were produced in each colour. All 20 units sit on Hermera wheels, finished in high gloss black.

The interior of the car features extensive carbon fibre, and full-grain Nappa leather, either embossed or stitched with the Greek Fret motif of the Versace Fashion House. Further trinkets accompanied the car, in the form of his-and-hers suitcases, suitcarriers, blue calfskin driving shoes, leather driving gloves, keychains and even Versace-designed men's and women's watches, all emblazoned with the Versace logo, and only sold to Murciélago 640 Versace owners.

MURCIÉLAGO LP640 VERSACE ROADSTER

There exists some confusion and controversy as to how many LP640 Versace Roadsters exist. There might be two or three, or even possibly four. What appears beyond debate, however, is that at least one factory-produced, Versace Greek Fret motif-adorned Roadster was built in conjunction with Versace.

bright orange called Arancio Atlas. Each side of this Roadster also carries two Arancio 650-4 decals, one low down on the side of the front bumper, and the other quite high up above the rear wing. The interior mirrors the exterior colour scheme, with Arancio Atlas highlighted E-Gear paddles and central transmission disc, and Arancio stitching on the seats and the steering wheel, all set within

Fashionistas among us will appreciate Versace's simple treatment of this limited edition LP640 Coupé, and note that it is both deliberate and impressively subtle. It is also exclusive, with only 20 units produced worldwide. Even less of the LP640 Roadster variant (inset) were made with production numbers arguably in single figures. An interesting amalgamation of two very exciting Italian brands.
Opposite page: The 650-4 Roadster is striking in appearance, with its flashes of bright orange.

MURCIÉLAGO LP650-4 ROADSTER

In 2009, Lamborghini released the last open top version of the Murciélago, the LP650-4 Roadster. Limited to just 50 units, these cars featured a matching two-tone interior and exterior. All the units were painted in a minimally glossy grey colour, which Lamborghini calls Grigio Telesto, but the upper surfaces of the front spoiler and the side sills, as well as the brake callipers and the leading edge of the variable airflow cooling system vents were painted in a

a black cabin. The 650-4 Roadster features an asymmetric cockpit design, with the driver's side of the cabin and the transmission tunnel being clothed in black Alcantara Nera, while the passenger's side is covered in smooth black Nero Perseus leather. The uprated V12 engine in this car produces an extra 10bhp compared to the LP640, taking the maximum power up to 641bhp, while maximum torque remains unchanged at 487lb-ft. Top speed is 205mph, and the dash to 62mph takes just 3.40 seconds.

MURCIÉLAGO LP670-4 SV CHINA LIMITED EDITION

The Murciélago LP670-4 SV China Limited Edition was tacit recognition by Lamborghini of China's status as the world's fastest growing consumer market, and the world's fastest growing major economy, with a growth rate that has averaged almost ten per cent over the last 30 years. It seems incongruous that one of the world's most extrovert, elitist and expensive car manufacturers should choose to produce a limited edition of an already-limited edition of its flagship car, exclusively for China: after all, China is nominally a country with a socialist market economy, and one that

the International Monetary Fund ranks as 78th on a per capita income basis, but that is how capitalism follows the market and the money.

Presented at the 2010 Beijing Motor Show, the China Limited Edition was restricted to ten units only, and was essentially unchanged compared to the standard LP670-4 SV. Every car is painted in a flat grey colour, with an orange central stripe flanked by black, running from the front bumper over the bonnet and onto the roof. A large SV logo down the rear flank and a large rear wing complete the exterior package, while the interior features a black

Alcantara cockpit with Arancio E-Gear paddles, Arancio seat and mat stitching, and an Arancio transmission tunnel disc.

The China edition reaches 62mph from standstill in 3.2 seconds, and has a top speed of 211mph, although the latter is reduced to 209mph when the large rear wing is fitted.

THE LAMBORGHINI MURCIÉLAGO R-GT

The introduction of the Murciélago R-GT racer at the 2003 IAA Frankfurt motorshow was unusual, as Lamborghini rarely participates in motor sports. Developed together with Audi Sport and Reiter Engineering, the R-GT is a factory-sanctioned racing car for privateers, and was built to meet FIA and ALMS specifications, and designed to compete in the 1100kg and below class. The R-GT is rear-wheel drive only; has had its engine capacity reduced to 5998cc, and carries air restrictors. In contrast to the standard Murciélago road car, the R-GT is lowered,;has a full-length protruding front spoiler, and sports a massive fixed rear wing. The gearing of the R-GT can be altered to suit individual race tracks, and the car's top speed and acceleration are therefore determined by the gear ratios chosen. For the 2010 season, Reiter Engineering developed an evolution of the R-GT called the LP670 R-SV, that uses a standard factory 670hp engine, and complies with the new 2010 FIA GT1 World Championship regulations.

Below: The R-GT as it appears in the official Lamborghini calendar, 'Designed by the right foot.' and also on display at Sant'Agata.

Left: Murciélago LP670-4 SV China Limited Edition.

THE AVENTADOR

The last Murciélago, an Arancio Atlas LP670-4 SV, rolled off the Sant'Agata production line in early November 2010, destined for a Swiss customer. A total of 4099 Murciélagos were built between 2001 and 2010.

Lamborghini's replacement flagship, the Aventador LP700-4, was unveiled in February 2011 at the Geneva Motor Show, and represented a fundamental shift away from the Countach, Diablo and Murciélago. While superficial similarities remained – including the wedge profile and a normally-aspirated V12 engine – Maurizio Reggiani, the Head of Lamborghini's R&D section, claimed that the incoming car was almost 100 per cent new, compared to the Murciélago. Indeed, the Aventador, known internally as model LB834, is the first production Lamborghini to feature a carbon fibre monocoque, and also has a totally new engine, transmission, four-wheel drive system, suspension set-up and cabin architecture. Even the vertically-opening doors are different, in that they now move laterally as they ascend.

The monocoque was developed with help from Boeing Aerospace and made in-house at Sant'Agata, weighing just 147.5kg. The brand-new engine, designated L539, is only the second V12 production road engine in Lamborghini's five-and-a-half decade history, and is completely different to the classic Bizzarrini V12 stalwart. This 6498cc, 60-degree V12, sequential multi-point injection unit produces 700hp (691bhp) at 8250rpm and 690Nm (509lb-ft) torque at 5500rpm. The new seven-speed ISR (independent shifting rods) Graziano gearbox is a single clutch robotised manual, and can change gears in just 50 milliseconds. There is a new Haldex IV all-wheel drive system with a greater rear-end bias compared to the Murciélago. The LP700-4 features Formula One-style suspension, with a sophisticated inboard pushrod spring and damper setup. The interior of the Aventador is stunningly beautiful, the stand-out aspects being the spectacular central transmission tunnel shroud with an array of controls, and the dashboard TFT LCD instruments that can be configured to show either a giant speedometer or a giant rev counter. The exterior bodywork is dominated by vast air vents behind each door.

Clockwise from top: Aventador Roadster at Cliveden, the day before Murciélago chassis 1564 was bought. The Aventador Roadster began production in 2013 – the carbon fibre roof panels are designed to be quick and easy to remove and stow in the front luggage space. The first Dealer delivery of Aventadors took place in 2011, as did the preview of the Aventador shown here in action at Goodwood.

An Aventador LP 750-4 SV, from 2017 with Dianthus lattice-style wheels and a revived front end, which incorporates a body-coloured splitter in front of the centre air intake. The overall view is a more aggressive stance, enhanced by more areas of contrasting black body work, finished in satin.

The Aventador has a kerbweight of 1575kg, which is 90kg less than its predecessor, and is claimed to use 20 per cent less fuel, and to produce 20 per cent less carbon dioxide emissions, than the out-going Murciélago. With its front-to-rear weight distribution of 43 per cent to 57 per cent, Lamborghini say that the LP700-4 can do the 0-100km/h (0-62mph) dash in 2.9 seconds, on the way to a 350km/h (217mph) top speed.

Murciélago SV alongside an Aventador and an even faster craft at a Lamborghini day at Brooklands in Surrey.

G-BBDG

THE FERRUCCIO LAMBORGHINI MUSEUM

The revamped Ferruccio Lamborghini Museum is a tribute and an act of love, from a son, in memory of his father.

Conceived and established by Tonino Lamborghini, the renowned designer and entrepreneur, this multifaceted museum traces Ferruccio Lamborghini's life through period photographs, personal artefacts, official documents, and of course the tractors that established Ferruccio's fortune, and the supercars that subsequently made the Lamborghini name world famous.

The museum is located a few kilometres from the centre of Bologna – on Strada Provinciale Galliera 319, near Argelato – in an area called the Tonino Lamborghini Forum, which was previously the Lamborghini Oleodinamica factory.

I first visited the first Ferruccio Lamborghini Museum on Saturday, 8th September 2001, at the time of the launch of the Murciélago. This museum was located in the old Lamborghini Calor factory near to Ferruccio's birthplace, in Renazzo di Cento. Needless to say, this was one of the highlights of the whole five-day-long trip, as was the subsequent visit to Cavaliere Ferruccio Lamborghini's tomb.

The new museum is much larger, at 9000 square metres, and is beautifully executed. The design of the museum was led by Tonino Lamborghini, who originally studied Political Science and Economy, and has a degree in Mechanical Engineering as well as recently being awarded a degree in Industrial Design by the University of Mumbai. His extensive design and business interests span articles as diverse as watches, mobile phones, sunglasses, perfumes, furniture, leather goods, coffee, boutique hotels, restaurants and lounges.

Opened in 2014, the Ferruccio Lamborghini Museum features, amongst many other things, Ferruccio's first Carioca tractor, other Lamborghini tractors dating up to the mid-1970s, and a reconstruction of Ferruccio's first office. Examples of Lamborghini burners, heaters and cooling systems are also on display, as well as a Class 1 offshore powerboat with Lamborghini engines, and a Lamborghini helicopter with dual controls.

For the motoring and history enthusiast, the museum houses the Fiat Barchetta with which Ferruccio Lamborghini competed in the 1948 Mille Miglia – and whose design and modification by Ferruccio have been detailed elsewhere in this book – as well as a second barchetta. For the supercar enthusiast, the Museum features the 350 GTV prototype, a 400GT, a Jarama, a Jalpa, a Urraco, an Espada, a Miura SV and a Countach.

The museum, which includes a bookshop and has facilities for hosting corporate events and business meetings, is open throughout the year. Guided tours, led by experts, help bring the exhibits to life. Further information can be found at www.museolamborghini.com.

Left inset: Plenty to see inside from Espada to Countach, an F1 power boat to a helicopter. Among the many exhibits are some original tractors, as well as Ferruccio Lamborghini's heating and cooling machinery. The brochure, pictured above, is a collectible in itself.

EXPERT OPINIONS

MIKE PULLEN

Within the United Kingdom, there is probably no single engineer better known for servicing and repairing Countachs than Mike Pullen. He has moved effortlessly from working on the Countach to the Diablo, and then on to the Murciélago. Old school in approach, his working motto is to repair and resurrect, rather than to replace. He puts this down to having been interested in all things mechanical since childhood – a self-confessed petrolhead, for whom the mechanics of watches and turntables are just as fascinating as the mechanics of his beloved cars and motorcycles. A serial Countach owner, he currently owns five; Mike says that his interest in Lamborghinis was sparked by their physical beauty and their rarity.

Mike Pullen works out of a small workshop in Haywards Heath, West Sussex. The size of his Carrera Sport workshop belies the genius within. At my last visit there, the garage was packed with cars, but even more astounding than the supercars were the disassembled engines and gearboxes, each pristinely clean and waiting to be reunited with their cars.

Mike says that all the different Murciélago variants are equally reliable. He feels that the Murciélagos are comparable in their reliability to their immediate predecessors, the Diablo and the Countach, and that the quality of assembly is better with the Murciélago.

Minor electrical issues are the main niggles with the Murciélago,

and the active rear spoiler ranks high here. The rods and brackets that raise and lower the variable airflow cooling system vents are also prone to breaking. Problems with the solenoids that operate the central locking system can cause the doors to repeatedly cycle through an unwanted sequence of locking and unlocking.

Gearbox issues are very rare, but there have apparently been a few – again, rare – issues with porous engine blocks in the 2002 cars. Splits and cracks are known to occur in the small water pipe that bleeds the engine's coolant system. Mike feels that the Murciélago, whatever its variant, is an inherently robust car.

The E-Gear system uses exactly the same gearbox and clutch as a manual transmission system, and is essentially reliable, though he feels that its single clutch system is slow. Bear in mind that we are talking to a track-day enthusiast here, who was known in his earlier days for really pushing his purple LP400 S Countach hard when on track. Resolving problems with the E-Gear actuators may well require engine removal, and the labour for this can be expensive.

The front lift system on the Murciélago rarely leaks, unlike the Diablo. Mike Pullen advises against getting drilled brake discs, as cracking around the holes is common. He favours replacement non-drilled Tarox brake discs.

Problems with any one of the 12 coil-over packs is not uncommon, whereupon the car will suffer a loss of power. The four electronic throttle bodies can also be troublesome, and expensive to correct. When they become defective, the car's engine revs can rush up to 2000rpm, and stay there without dropping back to idle, even when the engine has been slowly and sympathetically warmed up.

Rust is common on the square-tubed spaceframe chassis, the suspension components, and particularly where the steering rack bolts onto the chassis framework, due to the adverse electrolytic interaction between the adjacent steel and aluminium.

The thin steel roof of the Murciélago is susceptible to dents, particularly when drivers and passengers hold onto the roof for getting into, and out of the car. The carbon fibre bodywork is generally resilient, but a pattern can develop in the paintwork immediately above the radiators, due to the underlying heat. Misting up of the rear light clusters can be solved by drilling some tiny holes in the back of the cluster housing. Mike Pullen recommends that the Murciélago is kept on a battery charger, as low voltage from a failing battery can cause numerous, difficult to diagnose, problems.

Carrera Sport has not had any difficulty getting spare parts for the Murciélago from Sant'Agata thus far. Lamborghini-badged spare parts can be expensive, but cheaper, equally high quality, electrical components can often be obtained from Volkswagen and Audi, if the item codes are known.

ROBERTO GRIMALDI

Grimaldi Engineering, based in Halstead, Essex, in the United Kingdom, is an Italian-Swiss partnership, with Roberto Grimaldi both overseeing the workshop and very actively getting his hands oily, while the delightful Annabel Grimaldi fronts the firm.

Roberto has an impeccable engineering background, having served four years as an apprentice with the renowned Lamborghini specialist Colin Clarke, after which he continued with that company for a further eight years.

12 years ago, he and Annabel set up Grimaldi Engineering, and he has developed a well-deserved reputation for very high quality servicing and repair work on Italian exotica, particularly Lamborghini and Ferrari cars. He takes pride in training his juniors, and tries to impart to them his motto: "That will do,' won't do."

Roberto is of the opinion that the Murciélago is significantly more reliable than the later Diablo cars, principally because the electronic systems in the Murciélago are more modern and better developed. He points out that the Murciélago has more electronics, and while this in itself inevitably means more opportunities for glitches, Audi's quality control and greater resources more than compensates for this.

When asked specifically about the E-Gear system, Roberto says that this is a fundamentally robust system, which largely uses the same

gearbox mechanicals as the manual version. He says that it is very reliable if correctly set up. The Magnetti Marelli robotised system used in the Murciélago is, again, a variant of what had already been used by Ferrari and Maserati, so was not new at the time of its introduction in the 2005 Murciélago. However, the software that each company uses is different.

With regards to the E-Gear's reputation for excessive clutch wear, Grimaldi says that this is undeserved, but that the driver needs to treat the system with the same mechanical sympathy that one would a conventional manual system, and that the system needs to have been correctly set up.

He explains that the robotised manual system was originally developed for Formula One use, and that very little clutch wear actually occurs during gear changes at high RPM (high engine speeds). The majority of clutch wear occurs during the very short duration of clutch engagement at pull off (low engine speeds). Excessive wear occurs with excessive slippage of the single dry plate clutch, as seen in low speed city driving where stop-start driving is often unavoidable. He says that clutch life can be substantially extended by being decisive with the accelerator pedal, as slight, hesitant and frequent on-off applications of the throttle is akin to riding the clutch in heavy traffic. Clutch wear can also be minimised by not using the gearbox and the clutch as an accessory brake, so brake well before the corner, and get into the correct lower gear before entering the bend.

When asked about the Murciélago's electronic weak points, Roberto says that these are few in number. He highlights that the wipers can stop working in the rain, as the wiper ECU is located under the scuttle panel, and stops functioning when damp. He has a solution for this: he seals the ECU in a water-proof bag. Other weak points are door locks that falter in the wet (Roberto again recommends sealing the central locking ECU in a bag), and breakdown of the ECU that controls the rear spoiler and the movable VACS air vents.

Difficult starting and slow cranking on start-up may be due to degradation of the earthing straps from the engine to the chassis through rust. The rear earth cables are the most susceptible to rust, and Roberto always removes these cables, and thoroughly cleans them before replacing them, and adds on additional such cables if necessary.

Mechanical weak points are again apparently rare. One potentially catastrophic fault is the crank pulley bolt becoming loose, which will destroy the crankshaft. Roberto says that every Murciélago that comes through his workshop has this bolt inspected, and ratcheted to ensure adequate tightness. Using Locktite or Threadlock to secure the bolt might be a good idea, but the right type of adhesive needs to be used, and used sparingly and correctly, as the wrong or excessive adhesive can result in a bolt that is stuck in position.

Another potential mechanical weak point is the front drive shaft. If the car is often driven with full steering lock on, the constant velocity joints become damaged, and with it the front drive shafts. Grimaldi recommends never using the full steering lock, but says that being just off the maximum lock is fine.

Condensation within the Murciélago's rear light cluster can be tackled by drilling a small hole at the bottom of the backing plate, and condensation within the headlight covers can be minimised by ensuring that the back seals are intact and correctly seated.

Roberto says that the Murciélago's front lift system is generally robust, but that the car should not be left in the raised position for extended periods, as this places an unnecessary strain on the shock absorber seals. There is a remote solenoid switch incorporated within the front lift system, and if this fails the front lift system will fail to operate correctly. Replacing this solenoid switch is neither difficult nor expensive.

When asked about what expensive things go wrong with the Murciélago, Roberto instantly points to the crankshaft pulley bolt issue already discussed above. The car's differentials are also expensive, although it may be possible to rebuild them. ECUs can also be very expensive, and while some companies say that these can be repaired, Roberto says that in his experience, 80 per cent of ECUs sent for repair never get fully repaired.

Any repair that requires removal of the engine immediately becomes costly, just because of the time required to remove, and later replace, the engine. Engine removal requires draining of the coolant and the engine oil in this dry sumped car, and a long reach engine hoist is needed. Ideally one man needs to control the hoist, and two additional men are needed, one on either side of the engine, to ensure that the engine does not swing and damage the bodywork or scratch the paintwork, during its egress from and ingress into the engine cavity.

Removal of the engine is required for renewal of the clutch, for changing the manifolds and for starter motor replacement. Any repair work that requires full access to the gearbox, and any E-Gear actuator repairs (or more likely replacement) will also require removal of the engine.

Imminent clutch replacement may be signalled by the E-Gear system being reluctant to change onto the next higher gear at high RPM, or by the system selecting neutral rather than the next higher gear as called for by the driver.

Actuator problems are signalled by the E-Gear system not working properly, and starter motor problems are signalled by a clicking sound on turning the ignition key, with or without the starter motor whirring into life.

Roberto says that almost all Murciélago parts, be they body panels, mechanical parts or electronic components, are currently still readily available. Although some parts are clearly from Audi, for example the heater control system, most other parts are bespoke. Apparently Lamborghini and Audi have finally acknowledged that the 'Lamborghini Tax' is now a widely recognised and disliked thing, and have addressed this problem so that many of the Audi parts used in the Murciélago are available from Lamborghini at almost the same price as their identical Audi-labelled counterparts.

Right: Traction out of control: A four-minute test track drive that turned Simon's stalwart Murciélago into a four-year project of repairs, restoration and renewal.
Above: During happier days on the track.

SIMON GEORGE

Simon George is co-founder of 6th Gear Experience, a UK-based supercar driving experience company, and holds the distinction of having driven the majority of the 265,000 miles (425,000km) recorded on his Arancio Atlas 2004 Murciélago Coupé. He also owns a white LP640 Coupé, and his company has technicians capable of servicing and repairing Murciélagos. He is therefore uniquely positioned to comment on the driving characteristics and the maintenance of the Murciélago:

"On taking delivery of my new Murciélago way back in September 2004, I wouldn't have guessed that almost 14 years later it would have racked up enough mileage to get to the moon. SG54LAM became the jewel in the crown of newly-formed 6th Gear's fleet of supercars in 2006, and spent most weekends pounding around race circuits up and down the UK with paying members of the public at the wheel. The Murciélago quickly became a favourite with punters, and by 2010 had seen almost 8000 different drivers. This as well as being used daily on a three-hour round trip to the office.

So, mechanically, how has it been? Well, in a nutshell, pretty much bombproof. To date it has required eight clutches, three top end rebuilds and two sets of chains. I can't even guess how many sets of Pirellis the Murciélago has worn out over the years, but it is well into three figures.

The clutch wear has been interesting. I've lost count of the number of enthusiasts who seem to think they are weak. They really aren't. What one must never do with a Murciélago clutch is to slip it, or

attempt to reverse the heavy Lambo uphill. Both these actions guarantee premature wear.

Most of the faults that materialise over time are niggling: for example, the rear spoiler that becomes stuck in the up position; early rear light clusters becoming misted up, and hinges that raise the twin 'bat wings' snapping. More specifically, early Murciélago weak spots include temperamental throttle bodies that cause the car to 'hunt' at idle, corroding earth straps that cause the V12 to turn over extremely slowly, and, if you've ever fitted a sat nav or replacement sound system and then found that the fuel gauge won't budge from empty, strengthen the shielding on the wiring to it.

Pile on the miles – and I mean 50,000 (80,000km) plus – and if you're a careful driver you should still be on your second clutch, although a top end rebuild won't be far away.

Drive the Murciélago hard over many years, and you'll probably have the camshaft pulley let go. Mine sheared off at about 120,000 miles, and it's an expensive repair that obviously requires a new crankshaft. In my experience, this is the only major weakness with an ultra high mileage Murciélago.

From 2001 to 2005, Murciélagos had to live with the same brake set-up as on the Diablo. The brakes in early cars are just about adequate for the road, but quickly fade on the track. This doubtless led to the factory fitting the brake set-up from the Gallardo from 2005, which absolutely transformed the brakes. My 2004 car, unfortunately, just missed this upgrade, so I had some technicians retro-fit the front axle with the bigger set-up. This means that the pretty first generation wheels will no longer fit, so factor in a new set of Hemera or Hercules wheels (as fitted to post-2005 Murciélagos) if you are thinking of doing the same.

Moving onto exhausts, mine has had an abundance at different times of its life. The factory set-up doesn't sound particularly inspiring, so mine has sported several Tubi – and a Larini – systems. Sound is an acquired taste, of course, and right now my Murciélago is running with the factory back box, but Tubi style tips that generate a

The Lamborghini factory, September 2001.
Top: recently assembled pistons, awaiting engine blocks, to become the heart of the machine.
Centre: Freshly-painted doors during assembly on the factory floor.
Bottom: Yes, it's a really amazing place!

lovely sound without being too intrusive. Back in 2007 I experimented with removing the catalytic converters, fitting straight-through pipes with a Tubi style lightweight back box and tips. The result? Absolutely hellish to be honest! It was so loud the local council sent me a noise pollution notice, and the CD player rattled so much that it worked its way out of its mounting.

Calamity struck in 2012. On a wet test track with little run off and with a customer behind the wheel, the big Lambo exited a chicane at speed, spun, and hit an oak tree. No injuries to instructor or customer thankfully, but the Murciélago was utterly devastated. Apart from severe front end damage, the chassis was bent, the roof warped, and the gearbox and drive train scrap. An insurance claim would have meant being paid, of course, but in turn losing the Murciélago – and given its sentimentality, I decided to embark on what turned into a four-year project to completely rebuild the car from the ground up.

First job was the chassis, which was straightened by experts Chartwell in Derby, after which the Murciélago was trailered to Lamborghini Manchester to get it fixed.

Four years later, in early 2015, the V12 rumbled into life for the first time. After a complete new interior, gearbox, front end including headlights (at £5000), wheels, drive train, brakes and hundreds of other parts, the Lamborghini hit the tarmac once again.

My Murciélago is now retired from race circuits. That said, life hasn't become much easier for it, since I'm behind the wheel most days, in all weathers, taking the mileage ever closer to the 300,000 (480,000km) goal.

So what does the future hold for my Murciélago? Well given its pretty well-documented history and mileage, I'm guessing it's practically impossible to sell. On a positive note though, unlike possibly every other Murciélago owner, I don't have to worry about depreciation – so all things considered, I guess I'll be using it daily until I'm too old to get in and out of it!"

IAN HUNT

Photographs bring text to life, and I am immensely grateful to Ian Hunt and every other contributor who has given additional soul to this book through their images. Murciélagos captivate partly through their spectacular appearance, and Ian, a veteran photographer and life-long Lamborghini enthusiast, has captured them through his camera on three different continents. Here he tells us how:

"Ensure the vehicle is thoroughly clean after a wash and polish, and clear of dripping water. Tyre paint can add a sheen.

Lighting: Choose a bright, dry day with light high cloud to prevent harsh shadows. Add a polarising filter to minimise reflections from the car's glass, and to intensify blue skies.

Location: Select open areas. There should be nothing in the background or surroundings to distract from your main subject.

Positioning: The way natural light falls on the shapes, lines and details can make or mar the final results, so choose accordingly. Place the vehicle with the front wheels straight ahead within the arches.

Camera/lens: Careful choice of camera and lens to prevent distortion of shape and perspective. I use a lens range length between 24mm and 50mm on a full-frame DSLR camera. Start off with front three-quarter, with camera positioned at knee level, then similarly positioned camera rear three-quarter to ensure an overall view to capture the form. Move around the car to picture front, rear, side profile, then pick out strong or unusual design features, eg 'bat wing' air vents on the Murciélago. Use a tripod for maximum stability and clarity of focus. Altering the lens aperture will determine just how much depth of field we get, and determines how much the car versus the background will be in sharp focus or blurred.

Shooting a moving car from a static position by panning: Lower body remains still, while steady twisting of the upper body while holding the camera level and steady. Follow the car, and a press of the shutter button captures a sharp image of the car while blurring the background to give the visual impression of speed. Either 1/125th or 1/160th of a second.

The following are all considerations personal to the photographer:
·Film vs digital.
·Colour vs black and white.
·DSLR vs Compact vs Mobile Phone Camera.
·Storing images: Memory card capacity vs External Hard Drives which are capable of holding larger number of files.
·Transfer of Images: camera tethered by a USB cable vs memory card slot vs separate memory card readers via USB slot or cable. I have even emailed images from my camera phone to my email account.
·Photo Manipulation: Programmes such as iPhoto and others will allow basic adjustments to be made to exposure, contrast, sharpness, saturation and cropping."

ED BOLIAN, VIN & PRODUCTION NUMBERS

Ed Bolian has worked in the car business for the last 15 years, in traditional and entrepreneurial capacities. For six of those years, he was the Director of Sales at Lamborghini Atlanta. He now runs a social vehicle history reporting app called VINwiki that crowdsources the history of vehicles and allows enthusiasts to document their cars. He has owned six Lamborghinis, including four Murciélago LP640s, and currently daily drives a 2007 manual transmission LP640 that was originally delivered to Canada, and fraudulently reported stolen prior to his ownership by a dealer trying to avoid a charge for driving while drunk.

Ed writes the following:

"During the ten years the Murciélago was manufactured by Lamborghini (2001-2010), 4099 units were produced and sold. During a similar production run, Ferrari built 3083 550 Maranello units (1997-2001) and 2056 575 units (2001-2005). At the same time, they produced over 20,000 Ferrari V8 cars. The production of the Murciélago was weighted heavily towards 2007-2008, largely due to the rude health of the US economy during that period. This means that the early cars are extremely rare, generally with 250-400 cars built by hand per year.

While the styling of the car is as timeless as most vehicles designed and built by Lamborghini, the car's innards actually dated quickly. Its more modern counterparts have carbon tubs, sequential gearboxes, and much more advanced electronics. Since most collectors view these as maintenance liabilities and detractions from the experience of driving, the Murciélago stands as what will likely be the last pure expression of what a V12 Lamborghini supercar can be. As it replaced the iconic Diablo, it drew strongly from an uncompromising heritage, threading the line between Lamborghini's Italian passion and Audi's intervention.

The Gallardo was the first project truly managed by Audi. When Lamborghini was purchased in 1998, the Murciélago project was already generally defined. As we examine it today, the Murciélago and Gallardo have about as much in common as the Huracan and Urus do. They existed as very different cars. As a salesman in a Lamborghini dealership from 2009-2015, I was frequently asked by customers if they should upgrade from their Gallardos to a Murciélago. I explained to them that an ownership transition between the two cars was not an upgrade, but instead a departure. They are entirely different.

The Murciélago is big, brutal, and designed with a deliberately limited scope. Like most of its V12 lineage, the car was created in a world with very limited use of the word "no." It is crazy, it is uncompromising, it is excessive. The drama sparked when you see the car is consummated in the driving experience. I have owned four of these cars and I find them impossible to replace. Ten to 15 years after their creation, their capacity for emotional stirring is unmatched by anything engineered since. Adjectives are difficult to find to encapsulate the presence and the aura of a Murciélago, but "impressive" is the closest I have found. The car can stop you in your tracks and dominate the senses, even amongst a crowd of considerably more valuable cars.

As it entered a world that changed rapidly around it, the Murciélago found itself being compared to strange peers it never would have expected. The Ferrari Enzo has less than 100 more horsepower, but costs three times as

much. The Carrera GT and Mercedes McLaren SLR are nearly double the price. The price point of the Murciélago started around $280,000 in 2001, and topped off around $490,000 with the 2010 LP670-4 SV (ignoring the Reventón). It was built with a hypercar mentality, but priced like a sports car. The LP640 was most comparable, in 2006, to the newly-released $1.2 million Veyron, which in its street driving configuration could only manage 230mph, 19mph more than the LP640. The car was peerless at its price point throughout its production.

The price point of the Murciélago, combined with sufficient

production for them to be seen on the used market, created an interesting new paradigm for vehicle value in the USA. The Murciélago was the first pre-owned car with more than 1200 units produced (excluding the Ferrari F40 and Carrera GT) to retain a pre-owned value of more than $200,000 for several years. It created new challenges for banks, insurance companies, and dealerships that are now considered non-issues. Today, as these cars have been appreciating since 2013, the Murciélago market has once again become a dynamic place.

The Murciélago represents one of the last analogue hypercars. It has a fixed suspension, mechanical releases for all panels, limited electronic aids, minimal active aerodynamics, low specific output, an available manual gearbox, and a low-fi cockpit with simple instrumentation. It has a benign aftermarket radio and a logical engine management setup. It is a hypercar that should still work 50 years from now, which is not something we can take for granted with some of its more modern counterparts. It was never built to pursue every inch of performance imaginable. It was conceived to be the most timely and dramatic expression of the classic Lamborghini ideology, which has changed considerably since the Murciélago's demise. It seems likely that the Murciélago will remain the best car to define what the original Lamborghini ethos set out to be.

The Murciélago evolved considerably over its period of production. The first year's cars – model year 2002 in the USA – were fairly simple, with very few options. You could choose your interior colour, stitching, carpet, navigation, and a car cover to add to the base MSRP of $279,800. All of the 2002 and 2003 cars were built as six-speed manual cars. In 2003, very little changed apart from the introduction of the 50 Verde

Artemis (jade green) cars built to commemorate the 40th Anniversary of the brand. The early cars continued in the Diablo VIN convention, with prefixes of 'ZA9BC10' before a letter to designate country ('U' for USA, 'E' for Europe, etc). In the USA, this would be followed by a check digit and a '2' or '3' for model years 2002 and 2003, respectively. Then 'LA' and a serial number would follow.

During the 2003 model year in the US the VIN convention changed. 'ZHW' became the Lamborghini prefix, 'B' designated Murciélago, and the country code followed. Coupés, which were the only option until model year 2005, were designated '16' and the gearbox was defined by the next character. 'M' indicated manual, and 'S' indicated E-Gear. A check digit and model year indicator would follow, like all cars in the USA. So the 2003 and onward VIN convention resulted in a prefix of 'ZHWBU16M,' or, on cars for model year 2004 with the optional E-Gear sequential manual gearbox, 'ZHWBU16S.'

For 2005, the Roadster became available in the US. Rather than the designation '16' in the VIN for a Coupé, a Roadster had '26.' They were available as both manual and E-Gear cars. While the production ratio was not documented by the factory, it appears that fewer than 75 manual Roadsters came to the USA in '05-'06, compared to approximately 500 E-Gear cars. The 2005 and 2006 cars also had a different suspension set-up – which was much more reliable – as well as better instrumentation, larger brakes, wider seats, and a few more options, including a transparent engine bonnet.

The 2007 model year brought the LP640; truly a second generation of the Murciélago. It received a larger engine (6.5L over the previous 6.2), more power, more torque, better brakes, a more sophisticated interior, and new wheels and tires. Offered as both a Coupé or a Roadster from the start, the VIN would begin with 'ZHWBU37' for a Coupé and '47' for a Roadster. The 'M' or 'S' would follow to indicate manual or E-Gear.

2008 brought some carbon improvements and very little else to

2001	65 Coupés
2002	442 Coupés
2003	424 Coupés
2004	304 Coupés, 80 Roadsters
2005	230 Coupés, 234 Roadsters
2006	323 Coupés, 121 Roadsters
2007	423 Coupés, 206 Roadsters
2008	454 Coupés, 183 Roadsters
2009	274 Coupés, 57 Roadsters
2010	145 Coupés, 18 Roadsters
Total Coupés	3084
Total Roadsters	899

the LP640. The Reventón was produced as a limited run variant of the Murciélago. In 2009, ceramic brakes became standard, some buttons changed, a new nav system was used, and the E-Gear programming was noticeably improved.

2010 brought the LP670-4 SV variant, as well as 50 end-of-run LP650-4 Roadsters that used some SV components (steering wheels and mirror), had transparent engine bonnets, and a Grigio Telesto/Orange paint job. Lamborghini originally intended to produce 350 examples of the SV, but they did not sell well; the best estimate of total production is 186 cars worldwide, with 40 going to the USA. Their VINs begin in 'ZHWBU8AH.'

I manage a USA database of over 160 million cars, and I have

documentation on most of the Murciélagos that are in the USA to some extent. One of the first things that stands out as you peruse the histories of these cars is how many have been crashed. Due to the rear weight bias, the limited driver aids, and the value of the carbon fibre panels on each leading edge ($18k for a front bumper, $24k for a rear bumper before paint and install), they are easy cars to be deemed a total loss by insurance companies. I would venture to say that fewer than 50 per cent of the cars are still in original condition, and, as they get driven more, this flock gets smaller each day.

My favourite word to describe the Murciélago is 'useless.' It is a car that refuses to abide by any demands or requirements for an automobile to be functional. It is bulky, low, and has bad sight lines. The cabin is cramped, particularly for anyone over six feet tall. It doesn't fit many places, and the turning radius is bad. It is too low for many legally-constructed road and driveway transitions, so your destinations become limited. It makes too much power and goes too fast to be a normal car, but using it like one is one of the most amazing automotive experiences ever.

I have driven over 50,000 miles in Murciélagos, and no car makes any drive feel more special. There is a martyrdom complex to it, where you appreciate that it wasn't made to do whatever you are asking it to do, but strangely it acquiesces.

They have been inexpensive to maintain for me, and once you remove the child safety shelf from the front you can pack for two people for a week in the front trunk, and they are great cars for long trips. I have done several trips of 1000-2500 miles in Murciélagos, and they are brilliant. The sound, performance, and capability are fantastic, and the engagement sublime."

Ed Bolian has also published the following previously:

LAMBORGHINI MURCIÉLAGO PRODUCTION NUMBERS

"I came across a breakdown of year by year Production Numbers for the Murciélago on Wikipedia. The numbers were compiled from annual reports by Volkswagen. There were actually 4099 produced and this only totals to 3983:

So what do these numbers mean? They certainly demonstrate how rare Lamborghinis are. There are also some discrepancies between these numbers and common beliefs. For instance, Lamborghini said that they would produce 350 2010 LP670-4 SVs. They actually built approximately 186, making them even more rare than anticipated. There are approximately only 30 of them remaining in the USA.

It should also provide some guidance as to shopping. Let's say someone walks into our showroom and wants an '05-'06 brightly-coloured Coupé with under 10,000 miles.

That seems like a fairly open request. Let's see how many choices they should have.

Narrowing the scope to 2005 and 2006 Coupés would bring the field down to 353 cars. Approximately 20 per cent of the cars were brought to the USA. If they only wanted a USA car in one of the bright colours (yellow, orange, green) – which is a common request – then they would rule out at least 60 per cent of the 20 per cent imported, leaving around 28 cars eligible. Then, assuming ten per cent will have been in some kind of accident, they would have 25. If they wanted a car under 10,000 miles, with the average car being driven 2000-5000 miles per year, it might take out another 30 per cent, leaving 20. That, of course, would not be the number of cars on the market. With Lamborghinis, we typically see between five and ten per cent of the cars available for sale at any one time, meaning one or two cars would meet that specification. This means shopping takes time and patience. With Murciélagos worth between $100,000 and $450,000, buyers are sure to exclude certain cars based on cost, which will inevitably further limit choice. Shopping for an exotic car is a fun and exciting process, but it is something that should be entered into with open eyes and realistic expectations."

The Murciélago 40th Anniversary was produced in low numbers, with only 50 examples being built – all finished in Verde Artemis, a deep jade green/blue. The Murciélago Versace LP640 Coupé was limited to just 20 units having been produced. Apparently as few as 3 or 4 Versace Roadster units were made. See page 123.

RANDOM MUSINGS OF A DILETTANTE

To consider the Murciélago purely by its definition as a car would be an act of self-deprivation. The Murciélago can, and should, be appreciated on many different levels.

Of course it *is* a car, and the special experience of driving a Murciélago is the major part of what elevates it above more ordinary cars. However, its heritage – and particularly that of its spaceframe chassis and Bizzarrini engine, both of which can be aged, not in years, but in decades – is also very special. Its wedge-shaped forefathers, the Countach and Diablo, were the definitive supercars of their era, at a time when rival supercar manufacturers were fewer. Although less bespoke, less labour intensive to produce, and less rare than either the Countach or the Diablo, the Murciélago was still very much a hand-built car, in an era of mass production.

The dramatic angular exterior styling, which immediately catches the eye, is further enhanced by more subtle design details, such as the movable air-vents, the drama of the headlight covers, the integrated active rear spoiler, and of course the guillotine doors. The relatively restrained interior is luxurious, and a delight to behold.

V12 engines have always been exclusive, and multi-cylindered naturally-aspirated engines will become increasingly less common in a world conscious of climate change. The Murciélago was a product of its time, but – like every motorised vehicle since Flemish missionary/astronomer Ferdinand Verbiest's 1672 steam-powered ball-shaped boiler vehicle for the Kangxi Emperor of China, Nicolas-Joseph Cugnot's 1769 steam-powered tricycle for hauling artillery, or Carl Benz's 1886 three-wheeled Benz Patent-Motorwagen Nummer 1 – the Murciélago could not resist the march of progress. The Bizzarrini-designed V12 died with the Murciélago, and one day we might well see a flagship Lamborghini supercar powered by a lesser-cylindered hybrid engine with forced induction, or a purely electric-powered car. Meanwhile, we need to fully appreciate this venerable V12, while working examples still exist.

The rarity of the Murciélago also confers upon its temporary custodian the responsibility to drive, garage, service and repair the car with due respect. Only 4099 of these cars were produced, and not many of 2018's 7.6 billion people would eschew the opportunity of owning such an icon. Even in a post-petroleum future of 2100, I suspect that the Murciélago will still be lusted after by many of the 11.2 billion people estimated to be living at that time. There is a responsibility for preserving a piece of automotive history for future generations.

Respect is also due to the designers, engineers, production line workers and, yes, the financiers, whose collective expertise enabled the Murciélago to come into existence. The Murciélago's history carries within it, directly or indirectly, some of the most notable names in car building: Lamborghini, Bizzarrini, Marchesi, Dallara, Stanzani, Gandini, Bertone, Wallace, Alfieri, Balboni, Donckerwolke and Piech. If you believe in the ancient Hindu doctrines that suggest that repeated human incantations upon an inanimate object give that object some

semblance of life and humanity, what could be a better analogy than the dedication and love that the Sant'Agata workers lavished upon these four-wheeled deities, to make them the icons that they are.

So, the Murciélago should be appreciated and loved from a multifaceted perspective, and treated in accordance with its design. The reality is that the Murciélago was not really designed to be used as an everyday car – it's more usable than its predecessors, certainly, but an everyday car for the average owner? No.

There is an often-quoted saying that I intensely disagree with, which is applied to the Murciélago as it is to many other supercars: "It's a car, so just drive it." Supercars of this era are temperamental.

A Mazda, in a Lamborghini book? Absolutely. Like the Murciélago, the MX-5 is a pure driver's car, with a direction that is geared toward the fulfilment of feeling as one with the machine – an attitude rather than simply an object of desire. It's a lot of fun, too.

I know almost nothing about horses; but see many parallels with my sister's Argentinian dressage horse; these cars need to be awoken gently, coaxed into life, the vital fluids gradually warmed up (which might take 20 miles of driving). Once all these things have been done with due mechanical sympathy, these cars need to be opened up, and taken close to the red line in at least second gear, on each and every outing.

These cars need a regular exercise schedule, even in the winter, though not on salt-laden roads. Leaving them sitting idle on a battery tender for long periods risks the drying out of seals, and seizure of mechanical components. This is a particular challenge in places where roads need to be salted in the winter, as the spaceframe chassis is vulnerable to corrosion and salt spray will inevitably collect in remote corners, which cannot be reached for cleaning purposes. Each time these cars are woken, they need to be fully warmed up (I wait until the radiator fans start up, which in the winter can take up to 25 minutes), so that the heat within the drivetrain has a chance of evaporating

the water produced through the combustion process. It is a sobering thought to know that for every litre of petrol used in a car's engine, about one litre of water is produced – some of which will stay within the car's systems until it evaporates. There is probably nothing worse for a car than just starting it up and idling it for five minutes, halfway through its winter slumber.

The reality of "just driving" these cars is almost guaranteed to disappoint. To drive a Countach from Chesterfield Gardens to Dunraven Street, at Mayfair speeds, is to directly experience Dante's *Inferno* concurrently with losing Milton's paradise, and a Murciélago is not much better.

To enjoy – to actually *experience* – these cars, they need to be stretched. The Murciélago will give a good account of itself on track, but it is essentially unsuited for such work on a number of counts: it is too heavy, too wide, has body panels which are expensive to replace in case of an accident, produces relatively little downforce, and – most importantly – has too many driver aids. This last factor is very important, as electronic and mechanical aids act as a filter and barrier between car and driver. The series one Mazda MX5 is such a lovely car to drive both on the road and on track due to Toshihiko Hirai, the father of this iconic sportscar, taking the Japanese concept of Jinba ittai – "Unity of horse and rider" – as his guiding principle, with nothing allowed to come between the two. To achieve this, Hirai chose to keep the series one cars very simple, with power steering being just about the only driver aid.

One of the joys of modern track day driving is to experience genuine downforce. For a driver to actually experience downforce, not only does the car in question need to produce a significant amount of negative lift at a reasonable speed, but the amount of downforce also needs to be significant relative to the static weight of the car. In other words, the driver will more readily perceive the effect of downforce during cornering, when a given amount of downforce is produced in a light car, than when the same amount of downforce is produced in a heavy car. Here, the Murciélago again shows itself to be deficient as a modern track day tool. The LP640 Coupé, for example, only produces 70kg of negative lift at 125mph, while a KTM X-Bow R produces 200kg of downforce at the same speed. As the LP640 weighs 1665kg and the carbon monocoque X-Bow R weighs 790kg, a downforce of 70kg represents only a 4.2 per cent weight increase in the Lamborghini, while at the same 125mph, the X-Bow R is producing downforce equivalent to 25.3 per cent of its weight.

So, we have to accept that the Murciélago is, and was designed to be, a road car. Road use brings with it the restrictions of speed limits, traffic rules, crowded roads and less than perfectly smooth tarmac. To even begin to stretch a Murciélago, you therefore need to meticulously plan every trip, particularly its timing.

I believe that the single most important possession needed to maximise the enjoyment of driving a Murciélago, is a 50-pence alarm clock. When set for 4:30am in the summer, some of the previously mentioned restrictions become less restrictive, and the Murciélago can more fully express itself; though never, ever, *ever* break local road traffic laws – they are there for a reason.

Never buy a car as an investment – there are more efficient and less risky ways of securing, and growing capital. Most cars depreciate, and while rare supercars might depreciate less than ordinary cars, any perceived capital gain has to take into account the losses incurred through inflation, garaging, routine servicing and non-scheduled repairs. Supercars are a luxury, not a necessity, and as economic cycles are a fact of life, there will be times when supercars drop dramatically in value, maybe to the point of becoming unsaleable, because those who have the means to indulge in these toys are often the ones who are most exposed to economic downturns.

This probably applies even more pointedly to the current crop of supercars than it does to the Murciélago. Luddites who want a troublesome old Italian car featuring dated electronics and even more antiquated mechanicals have less choice open to them than those who are open to buying a newer supercar. Lamborghini, Ferrari, McLaren and Porsche are all producing more models, and greater volumes of each model, than was the case previously. While it is true that the new wealth found in emerging markets can soak up this extra supply, part of the value of supercars lies in their rarity. In a globalised world, where cars can easily be advertised on the internet and cheaply shipped (or flown) across continents, excess supply equals poor residuals. My conclusion is that supercars are unreliable investments. Only buy them because you genuinely want them – whether to drive, gaze upon, or as subjects of a book.

If you must buy, then do your homework thoroughly. Be honest with yourself as to why you want a particular brand or type of car, and if it will truly meet your requirements and budget. Parts of the internet, specifically owner and enthusiast forums, are a rich font of knowledge. Attending car shows or marque-specific car club events might allow

Porsche 981 Boxster: 20inch titanium wheels, nitrous oxide injection, gold teeth and hybrid technology as standard, of course ... court calls, then.

a potential owner to examine the car of their dreams in detail, and owners who attend these gatherings are usually only too keen to talk for hours on end about their babies. Search out and talk also to specialist mechanics, who will know the strengths and foibles of each model, and who can guide one towards or away from a particular model variant. These are not insignificant purchases for the average enthusiast, so I usually aim to know more about the specific model variant that I have decided upon than the salesperson, who has the disadvantage of having to know something about every car in their showroom.

Buying directly from a private individual carries the benefit of meeting the seller, and getting a feeling for how he might have treated the car during his ownership tenure. I bought my first Ferrari – a 355 GTB – independently, and within a few minutes of meeting the very knowledgeable and straight-talking owner, it was obvious that he cherished the car. He insisted that only he would drive the car during the test drive, and the gentle way in which he warmed up the car spoke volumes. The gentleman from whom I bought my KTM X-Bow R had used it for one trip, and knew next to nothing about this extremely low-mileage car, but had 37 pristine cars in his collection, all of which lived within protective car cocoons, and every one of which he had serviced by the book. Meeting and gently interrogating owners can be very revealing.

Independent dealers can, like most things in life, be excellent or terrible, and often you will not find out about the quality of their after-care service until it is too late. Personally I have had superb service from every independent that I have bought from, bar one trader. Needless to say there are lessons that I learnt from this one experience, which might help others. Traders who do not know the basics about their stock should ring alarm bells. A trader who repeatedly insists, including in email, that Murciélagos have 20-inch wheels as standard should be treated with suspicion – they all left the factory with 18-inch wheels, even the LP670-4 and the Reventón. A trader who claims that Ferrari 355s have 12 cylinders, that KTM X-Bow Rs have steel spaceframe chassis, or that Porsche 981 Boxsters came with titanium wheels as standard, should be viewed with equal scepticism. Traders who, on a test drive, thrash cars from cold are probably not the best people to buy from. Examine paperwork carefully, and do challenge any discrepancy, even if the trader has further enticed you by leaving your long-awaited purchase idling outside his office. My sister, a Court of Appeal judge and   in a previous incarnation   a commercial barrister, has a long-standing mantra of avoiding the courtroom if at all possible. However, there are firm and clear laws that govern trade transactions in the United Kingdom, the European Union, the United States of America, and just about every country in the world, and these laws apply just as stringently to independent car traders as they do to any other commercial transaction. When negotiation fails to bring about a fair and amicable settlement, then the courts are the only remaining option. Success is never guaranteed, even in what may appear to be the most straightforward of cases, and the plaintiff has to carry a financial burden, at least initially. Only ever go to court if, after careful and dispassionate analysis, you genuinely feel that

GARAGE AT A GLANCE

you have a strong case. Help your lawyer by assembling all relevant case material in chronological order, particularly all electronic and hard copy correspondence. If you choose to present your own defence, as I did, learn the basics of the law (which is an interesting subject in itself, anyway), and be fully prepared. The courts usually judge in a fair manner, consistent with the prevailing law. Murciélagos, like just about every other supercar, induce a sense of astonished euphoria and consequent carelessness in otherwise calculating purchasers, and it is important to keep a sense of reality and be able to look past the shiny bodywork throughout the buying process. It is also important to reiterate that every other independent dealer that I have bought from has been exemplary, and one case should not tar a whole sub-industry.

Factory-franchised main dealers not only have their own reputations to maintain, but also that of their associated manufacturers. They are sometimes better equipped to appraise second-hand cars than independents (be it through having specialist mechanics, extensive garage facilities, or access to bespoke electronic interrogation systems, like LARA in the case of Lamborghini), but this is not always the case. Some independent dealers have invested heavily in personnel and garage equipment, and are more knowledgeable about older models than main dealers, who might only be used to working on the latest cars. Dealers with their own garage facilities carry a distinct advantage, both in being able to better appraise a car, and also in being better positioned to offer good after-sales care. Main dealer warranties are usually more comprehensive than the minimum level set by law.

When buying any secondhand car, and particularly one as complex as a Murciélago, it is money very well spent to arrange a comprehensive pre-purchase inspection by a trusted, reputable specialist mechanic, who has access to the necessary specialist mechanical and electronic tools. I cannot emphasise just how important this is.

Wherever and whomever you purchase from, buy a car on its present condition and its documented past history. The vendor is almost irrelevant – you are buying the car not the vendor.

Maintenance is a major issue that each prospective owner needs to have researched deeply before committing themselves to a purchase. Obviously, the cost of routine servicing, as well as the cost of unexpected repairs once outside the warranty period, need to be within the available budget. More challenging still can be finding a good mechanic. In an era of "technicians who replace parts, often in a random sequence," finding an engineer-mechanic is like finding the mother lode. Fellow owners and marque clubs will be able to advise on who has the necessary experience and tools to service a given model, but I always make a trip to meet and personally appraise the actual mechanic who will be working on my car, well before making a service booking. It can do no harm for the mechanic to know that the owner cares and is well-informed. Once you have found a good engineer-mechanic, don't let them go. Like a physician, a specialist mechanic who looks after a car over a long period will develop a deep knowledge of – and a bond with – that car, and will be able to exercise preventative maintenance. My Countach has only been serviced by three different people since it left Sant'Agata, more than 30

KTM's X-Bow R: outrageously styled, no frills driving for the road, and for the track. Not for the faint-hearted. An umbrella is de rigueur, especially in the UK.

Mazda MX-5 roadster: (tucked in at the rear) trusty, brilliantly executed design and balanced, driver oriented.

Porsche 981 Boxster S: Reliable, fun, powerful, yet every day driveable! German engineering and flair within a timeless body shape.

Lamborghini Countach LP5000 QV: All-time classic Lamborghini – almost defines the marque on its own. Future proof, – it will still turn heads and drop jaws in 50 years time.

What this book is all about: The Murciélago will follow in the Countach's footsteps as far as longevity goes. The car that marked collaboration, modernity and a major turning point for the Italian stable, and sent it off into its next, no doubt incredible, chapter. What next for Lamborghini?

Leaving it to the experts: HR Owen's well-equipped workshops, circa 2010.

years ago, and I am deeply grateful to Lee Cunningham, Mike Pullen and Roberto Grimaldi for their expertise. I have particular respect for those owners who have the knowledge and confidence to carry out service and repair work on their cars, but with Murciélagos, as with most other modern supercars, access to specialist tools and electronic software maps can be a major stumbling block.

Patina versus restoration is a debate that has no correct answer, and each enthusiast or owner will have an opinion that will be valid for their way of thinking. My personal approach to my cars is akin to the way I first approached my wife: be clear about what you want and why, appraise very carefully before purchase, buy the very best available, maintain by the book, and after that only do the minimum needed to keep them functional. Cosmetic surgery is unnecessary, but its mechanical equivalent may become necessary, on occasion.

I tend to give my cars a quick wash after each long drive, and,

with the original paint on a Murciélago being relatively soft, it is not uncommon to find a new pin prick-sized chip on the nose or bumper. While it hurts me to find it, this tiny chip is also a testament about a much enjoyed drive – a part of the car's history. Often I can remember which chip came into being on which drive, and to do a regular front-end respray purely to make an old car look like a new car is to rob it of its chance to grow old gracefully. Instead, I touch up each chip with a small sewing needle, which leaves a small blob as a reminder, but removes the ugly underlying primer showing through. A genuine hand painted job if ever there was one.

The provenance of a car is based on its authenticity, and unnecessary restoration undermines both. This applies equally to a wife, as it does to a car.

It may be wise to close at this point, while again stating that these are just some casual ruminations from an enthusiastic amateur.

LP670-4 SV sporting an appropriately designated Italy/Bologna Number plate, 'does the hill' during the Chomondeley Festival near Crewe, Cheshire – while a mesmerised crowd look on.

Spoiler Alert ...

The Book of the Ferrari 288 GTO
Joe Sackey

Covers the background, conception, design, production and aftermath of the iconic Ferrari 288 GTO, including the prototypes, the early production cars, the mainstream production cars in their various specification guises, and the Evolution cars planned for the aborted Group B FIA race series. A comprehensive and beautifully illustrated look at a unique sports car.

ISBN: 978-1-845842-73-4

The Book of the Lamborghini Urraco
Arnstein Landsem

This book tells the amazing story of the Lamborghini Urraco. It describes the problems that beset this little supercar, and why it never got the chance it truly deserved. After its demise, the Urraco lived on in the form of the Silhouette and the Jalpa, and both these models are also covered in detail in the pages of this book. Featuring detailed advice for potential buyers, as well as over 300 photos and illustrations, this is a fascinating and practical account of a supercar classic.

ISBN: 978-1-845842-86-4

– a few of our Italian classics

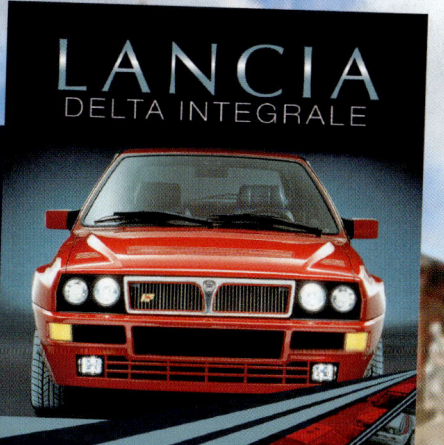

The art of captivation

INDEX

"This was the perfect moment when I